AF411347

Conservation Genetics for Management of Threatened Plant and Animal Species

Editors

Kym Ottewell
Margaret Byrne

MDPI • Basel • Beijing • Wuhan • Barcelona • Belgrade • Manchester • Tokyo • Cluj • Tianjin

Editors
Kym Ottewell
Conservation and Attractions
Australia

Margaret Byrne
Conservation and Attractions
Australia

Editorial Office
MDPI
St. Alban-Anlage 66
4052 Basel, Switzerland

This is a reprint of articles from the Special Issue published online in the open access journal *Diversity* (ISSN 1424-2818) (available at: https://www.mdpi.com/journal/diversity/special_issues/ threatened_plant_animal).

For citation purposes, cite each article independently as indicated on the article page online and as indicated below:

LastName, A.A.; LastName, B.B.; LastName, C.C. Article Title. *Journal Name* **Year**, *Volume Number*, Page Range.

ISBN 978-3-0365-4441-0 (Hbk)
ISBN 978-3-0365-4442-7 (PDF)

Cover image courtesy of Rujiporn Thavornkanlapachai

Contents

About the Editors vii

Preface to "Conservation Genetics for Management of Threatened Plant and Animal Species"   ix

Kym Ottewell and Margaret Byrne
Conservation Genetics for Management of Threatened Plant and Animal Species
Reprinted from: *Diversity* **2022**, *14*, 251, doi:10.3390/d14040251   1

Jane Sampson and Margaret Byrne
Genetic Differentiation among Subspecies of *Banksia nivea* (Proteaceae) Associated with Expansion and Habitat Specialization
Reprinted from: *Diversity* **2022**, *14*, 98, doi:10.3390/d14020098   5

Kor-jent Van Dijk, Michelle Waycott, Joe Quarmby, Doug Bickerton, Andrew H. Thornhill, Hugh Cross and Edward Biffin
Genomic Screening Reveals That the Endangered *Eucalyptus paludicola* (Myrtaceae) Is a Hybrid
Reprinted from: *Diversity* **2020**, *12*, 468, doi:10.3390/d12120468   19

Michael D. Amor, Neville G. Walsh and Elizabeth A. James
Genetic Patterns and Climate Modelling Reveal Challenges for Conserving *Sclerolaena napiformis* (Amaranthaceae s.l.) an Endemic Chenopod of Southeast Australia
Reprinted from: *Diversity* **2020**, *12*, 417, doi:10.3390/d12110417   33

Rujiporn Thavornkanlapachai, Esther Levy, You Li, Steven J. B. Cooper, Margaret Byrne and Kym Ottewell
Disentangling the Genetic Relationships of Three Closely Related Bandicoot Species across Southern and Western Australia
Reprinted from: *Diversity* **2020**, *13*, 2, doi:10.3390/d13010002   55

Daniel J. White, Kym Ottewell, Peter B. S. Spencer, Michael Smith, Jeff Short, Colleen Sims and Nicola J. Mitchell
Genetic Consequences of Multiple Translocations of the Banded Hare-Wallaby in Western Australia
Reprinted from: *Diversity* **2020**, *12*, 448, doi:10.3390/d12120448   75

Joseph P. Zilko, Dan Harley, Alexandra Pavlova and Paul Sunnucks
Applying Population Viability Analysis to Inform Genetic Rescue That Preserves Locally Unique Genetic Variation in a Critically Endangered Mammal
Reprinted from: *Diversity* **2021**, *13*, 382, doi:10.3390/d13080382   97

Patricia Escalante-Pliego, Noemí Matías-Ferrer, Patricia Rosas-Escobar, Gabriela Lara-Martínez, Karol Sepúlveda-González and Rodolfo Raigoza-Figueras
Genetic Diversity and Population Structure of Mesoamerican Scarlet Macaws in an Ex Situ Breeding Population in Mexico
Reprinted from: *Diversity* **2022**, *14*, 54, doi:10.3390/d14010054 119

Gael L. Glassock, Catherine E. Grueber, Katherine Belov and Carolyn J. Hogg
Reducing the Extinction Risk of Populations Threatened by Infectious Diseases
Reprinted from: *Diversity* **2021**, *13*, 63, doi:10.3390/d13020063 131

Francesco Palmas, Tommaso Righi, Alessio Musu, Cheoma Frongia, Cinzia Podda,
Melissa Serra, Andrea Splendiani, Vincenzo Caputo Barucchi and Andrea Sabatini
Pug-Headedness Anomaly in a Wild and Isolated Population of Native Mediterranean Trout
Salmo trutta L., 1758 Complex (Osteichthyes: Salmonidae)
Reprinted from: *Diversity* **2020**, *12*, 353, doi:10.3390/d12090353 . . . . **139**

About the Editors

Kym Ottewell

Kym Ottewell (Dr) is a Senior Research Scientist in the Department of Biodiversity, Conservation and Attractions, Western Australia. Her work focuses on the application of genetic research to conservation of threatened species, with a particular focus on mammals. Her work encompasses temperate and arid landscapes, with current projects covering a range of threatened marsupial, rodent and bat species investigating natural and anthropogenic impacts on genetic connectivity and genetic diversity.

Margaret Byrne

Margaret Byrne (Dr) is Executive Director, Biodiversity and Conservation Science in the Western Australian Department of Biodiversity, Conservation and Attractions where she leads a strong science group providing an evidence based approach to conservation management and policy. Her work informs biodiversity conservation strategies for management of landscapes and of rare and threatened species, and her phylogeographic studies have provided a greater understanding of the evolutionary history of the Australian biota, and its influence on current distributions, patterns of genetic diversity and location of refugia. Her current research is focused on applications of genomics in plant conservation and climate adaptation strategies.

Preface to "Conservation Genetics for Management of Threatened Plant and Animal Species"

Genetic diversity is fundamental to the maintenance of species diversity and ecosystem resilience, and especially the capacity of species to adapt to changing and challenging environmental conditions. Globally, species and ecosystems continue to decline as more species are added to threatened species lists every year. The field of conservation genetics offers a range of techniques and statistical approaches that enable us to describe and monitor various aspects of genetic diversity and make inferences about the underlying ecological and evolutionary processes driving these patterns in threatened species. While conservation genetic analytical approaches are sophisticated and well-advanced, there is now growing interest from managers to incorporate management of genetic diversity into conservation programs.

In this book, we highlight conservation genetic (and genomic) papers that demonstrate applied outcomes that inform practical threatened species management. We cover a broad range of species and genetic approaches, but focus on how conservation genetic information is used to underpin management actions for species recovery. Through the exposition of a diversity of approaches, we aim to demonstrate to conservation managers and researchers how conservation genetics can inform on-ground species management.

We cover topics including:

- Species and subspecies delimitation, and identification of management unit;
- Distribution of genetic diversity across populations;
- Using admixture for conservation purposes;
- Genetic erosion;
- Managing inbreeding in small populations;
- Hybridisation;
- Reintroductions/restoration using genetic principles;
- Genetic management for disease.

Kym Ottewell and Margaret Byrne

Editors

Editorial

Conservation Genetics for Management of Threatened Plant and Animal Species

Kym Ottewell * and Margaret Byrne *

Biodiversity and Conservation Science, Department of Biodiversity, Conservation and Attractions, Locked Bag 104, Bentley Delivery Centre, Perth, WA 6983, Australia
* Correspondence: kym.ottewell@dbca.wa.gov.au (K.O.); margaret.byrne@dbca.wa.gov.au (M.B.)

Citation: Ottewell, K.; Byrne, M. Conservation Genetics for Management of Threatened Plant and Animal Species. *Diversity* **2022**, *14*, 251. https://doi.org/10.3390/d14040251

Received: 14 March 2022
Accepted: 23 March 2022
Published: 28 March 2022

Publisher's Note: MDPI stays neutral with regard to jurisdictional claims in published maps and institutional affiliations.

1. Introduction

Globally, species and ecosystems continue to decline, and the impact on threatened species is increasing. The ongoing loss of intraspecific genetic diversity is contributing to the erosion of species' adaptive potential and can hasten population declines, especially in the face of increasing ecological and anthropogenic disturbance [1]. The field of conservation genetics offers a range of techniques and statistical approaches that enable us to describe and monitor various aspects of genetic diversity and make inferences about the underlying ecological and evolutionary processes driving these patterns, in turn, informing approaches to conservation management [2–4]. In addition, given the strong empirical support for the relationship between small population size, reduced genetic diversity and reduced population fitness [5], it is clear that ongoing monitoring of the genetic diversity in threatened species is going to be crucial for sustaining species' health into the future. Research in this field has progressed enormously in recent years, alongside growing knowledge on the application of genetics in conservation and interest from managers in incorporating the management of genetic diversity into conservation programs.

There are many areas of conservation where genetic data can provide direct information to guide management actions for threatened species. In this Special Issue, we highlight conservation genetic studies that demonstrate applied outcomes that inform practical threatened species management. These studies present a diversity of research approaches, both in genetic methodology (microsatellites, chloroplast and mitochondrial DNA sequencing, single nucleotide polymorphisms) and in combination with complementary statistical approaches, taken from allied disciplines (e.g., population viability analysis, climate niche modelling), with application to specific questions relating to on-the-ground species management.

1.1. Identification of Conservation Units

Fundamentally, species are the primary units of conservation and delineation of species and subspecies boundaries, and the identification of management units within species, is important to ensure appropriate conservation listings are in place and that conservation actions are properly targeted. This is highlighted in a study of *Banksia nivea*, where Sampson and Byrne [6] found that genetic relationships among subspecies were consistent with the existing taxonomy for two subspecies (one common, one endangered), but not for a third (previously considered rare), indicating further taxonomic assessment is required for *B. nivea* subsp. Morangup. Phylogenetic analyses revealed evidence for a more recent divergence of the localised subspecies, associated with expansion from the dryer sandy soils inhabited by the widespread subspecies into the winter-wet ironstone soils in the southwest of Western Australia, consistent with progressive long-term climatic drying.

Understanding species delimitation is also critical to identifying hybridisation and putative taxa of hybrid origin, with a view to assessing their conservation value. A study by van Dijik et al. [7] investigated the status of a conservation-listed tree that is restricted to the

Fleurieu Peninsula and Kangaroo Island of South Australia and suspected to be of hybrid origin. Genetic analysis of *Eucalyptus paludicola* and its putative parental species identified two genetically distinct clusters, comprising *E. ovata* and *E. cosmophylla*, while *E. paludicola* individuals were admixed between these two species, consistent with a hybrid origin. Given hybrid class assignment tests indicate that the majority of *E. paludicola* individuals are F1 hybrids with a low incidence of backcrossing, these data support the hypothesis that *E. paludicola* is a transient hybrid entity rather than a distinct hybrid species. As such, the authors find little support for the ongoing conservation listing of *E. paludicola* or, indeed, recognition as a distinct species.

1.2. Evaluating Population Structure and Genetic Diversity to Inform Management Approaches

Dispersal is important for maintaining genetic connectivity amongst populations and, consequently, understanding patterns of gene flow and genetic differentiation is critical if managers are to use admixture to maximise diversity and adaptive potential in reintroductions and restoration projects. This is explored by Amor et al. [8], who used a genetic analysis to investigate the genetic relationship among disjunct groups of remnant populations of *Sclerolaena napiformis*, a perennial chenopod endemic to southeast Australia. They found genetic differentiation among the three regions, with low genetic diversity within populations and high levels of inbreeding. A decline in abundance through habitat fragmentation is compounded by climate modelling that predicts a reduction in suitable habitat for the species, under even the most conservative climate change scenario. The study shows the benefits of applying the knowledge of genetic diversity in restoration and recommended an admixed provenance approach to souring of seed for restoration, both within and across regions, to maximise genetic diversity and maintain dynamic evolutionary processes driven by individual plant fitness in response to the novel environmental conditions.

Assessment of gene flow and genetic diversity amongst remnant populations can assist in focusing conservation attention to those most in need. Thavornkanlapachai et al. [9] provided an example of this in an investigation into the genetic relationships amongst a complex of closely related bandicoot species (genus *Isoodon*), which are variously threatened by ongoing habitat loss and predation by introduced predators. Analysis of mtDNA identified three major clades that largely resolved existing taxa, although, with a pattern of 'intermediate polyphyly' [10] observed between South Australian (SA) and Western Australian (WA) populations. This highlighted ancestral connections between these groups that were resolved as distinct entities in analyses with nuclear markers. SA and Victorian populations of *Isoodon* bandicoots were identified as suffering genetic erosion, emphasising the prioritisation of these populations for specific conservation efforts to reduce further loss of genetic diversity.

1.3. Managing Genetic Diversity in Translocations

Genetic analysis provides a basis to guide translocations and inform options for undertaking genetic rescue. White et al. [11] demonstrated how genetic data can be used in an investigation into the genetic effects of past translocations in a once widespread mammal species that is now restricted to islands and fenced enclosures in Australia, to inform future translocations of the species. Genetic analysis of the banded hare-wallaby (*Lagostrophus fasciatus*) showed serial translocation from a single source population has led to a loss of genetic diversity in a translocated fenced reserve, whilst inbreeding is of concern in the translocated island population. Population viability analysis and gene retention modelling indicated founder population sizes of ~100 individuals and mixing of two source populations were optimal to maximise demographic resilience and genetic richness in a planned translocation to facilitate persistence in the face of various stochastic environmental events.

1.4. Implementing Genetic Rescue to Manage Inbreeding

The introduction of new genetic diversity into an inbred population to assist in genetic rescue must be balanced against the risk of 'genetic swamping', leading to the loss or disruption of local adaptive genotypes. Zilko et al. [12] combined genetic data with population viability analysis to guide genetic rescue in the inbred lowland population of Leadbeater's possum (*Gymnobelideus leadbeateri*), a threatened species in the forest of south-eastern Australia, whilst balancing the maintenance of local adaptation. Translocated animals were sourced from the outbred and genetically diverse highland population and this resulted in higher retention of local alleles in the supplemented lowland population due to the reduction in genetic drift. Nevertheless, carrying capacity in the lowland population is currently insufficient to enable population recovery. Consequently, the authors recommended the establishment of a new population of lowland possums, in a high-quality habitat, and gene exchange with highland populations, to alleviate inbreeding depression and maximise the retention of locally unique neutral genetic variation.

1.5. Managing Genetic Diversity in Captive Breeding Populations

Captive breeding populations can be important sources of diversity to support reintroduction efforts for threatened animals, although ongoing assessment is required to ensure diversity is captured in the founding populations and breeding is managed to maximise retention of diversity. In a study of the captive Mesoamerican scarlet macaw (*Ara macao cyanoptera*) population at Xcaret Park, Escalante-Pliego et al. [13] revealed that founding and current breeders showed high retention of genetic diversity, low inbreeding and low relatedness, and that the captive population has a similar level of genetic diversity to the wild population in the Mayan Forest. As a consequence, the captive breeding population at Xcaret is an important source of birds for the reintroduction program of this subspecies.

1.6. Genetic Approaches to Managing Wildlife Disease

Managing in-situ populations exposed to disease threats, based on genetic principles, is an integral component of recovery programs for highly vulnerable species. Glassock et al. [14] reviewed the underlying principles of genetic management and highlighted how managing gene flow, diversity and inbreeding assists in reducing the extinction risk of Tasmanian devil (*Sarcophilus harrisii*) populations, threatened by the deadly devil facial tumour disease. This disease poses a significant risk to the persistence of the species. The supplementation of populations establishes gene flow and genetic analysis and can inform how best to increase adaptive potential, whilst minimising any potential for outbreeding depression or loss of local adaptation. This review provides timely discussion on the issues faced by conservation managers for managing emerging diseases in threatened species, where disease eradication is not possible.

Similarly, Palmas et al. [15] used genetic analyses to investigate whether there was an underlying genetic cause of a reported record of pug-headedness in a critically endangered population of native Mediterranean trout *Salmo trutta* from Sardinia, Italy. The genetic analysis suggested that inbreeding or outbreeding depression are not contributing factors in the instance of the deformity in this population, and variation in environmental factors during larval development seemed the most likely factors influencing the deformity.

2. Conclusions

These studies are just a few examples of the ways in which genetic analysis can inform effective conservation actions for the management of threatened species. Importantly, in these exemplar papers, authors provide explicit recommendations to enable positive impact on the management of their species of interest. We hope that conservation managers not so familiar with genetic tools and techniques find value in the studies provided here, in demonstrating the link between the conservation questions of interest, analytical approaches and the ensuing conservation management outcomes. Further information and exploration of the application of genetics and genomics for the conservation of threatened species is available in the book Conservation and the Genomics of Populations [16].

Author Contributions: Writing—original draft preparation, M.B. and K.O.; writing—review and editing, K.O. and M.B. All authors have read and agreed to the published version of the manuscript.

Funding: This research received no external funding.

Conflicts of Interest: The authors declare no conflict of interest.

References

1. Banks, S.C.; Cary, G.J.; Smith, A.L.; Davies, I.D.; Driscoll, D.A.; Gill, A.M.; Lindenmayer, D.B.; Peakall, R. How does ecological disturbance influence genetic diversity? *Trends Ecol. Evol.* **2013**, *28*, 670–679. [CrossRef] [PubMed]
2. Hohenlohe, P.A.; Funk, W.C.; Rajora, O.P. Population genomics for wildlife conservation and management. *Mol. Ecol.* **2021**, *30*, 62–82. [CrossRef] [PubMed]
3. Holderegger, R.; Balkenhol, N.; Bolliger, J.; Engler, J.O.; Gugerli, F.; Hochkirch, A.; Nowak, C.; Segelbacher, G.; Widmer, A.; Zachos, F.E. Conservation genetics: Linking science with practice. *Mol. Ecol.* **2019**, *28*, 3848–3856. [CrossRef] [PubMed]
4. Schwartz, M.K.; Luikart, G.; Waples, R.S. Genetic monitoring as a promising tool for conservation and management. *Trends Ecol. Evol.* **2007**, *22*, 25–33. [CrossRef] [PubMed]
5. Willi, Y.; Kristensen, T.N.; Sgro, C.M.; Weeks, A.R.; Orsted, M.; Hoffmann, A.A. Conservation genetics as a management tool: The five best-supported paradigms to assist the management of threatened species. *Proc. Natl. Acad. Sci. USA* **2022**, *119*, e2105076119. [CrossRef] [PubMed]
6. Sampson, J.; Byrne, M. Genetic differentiation among subspecies of *Banksia nivea* (Proteaceae) associated with expansion and habitat specialization. *Diversity* **2022**, *14*, 98. [CrossRef]
7. Van Dijk, K.-J.; Waycott, M.; Quarmby, J.; Bickerton, D.; Thornhill, A.H.; Cross, H.; Biffin, E. Genomic screening reveals that the Endangered *Eucalyptus paludicola* (Myrtaceae) is a hybrid. *Diversity* **2020**, *12*, 468. [CrossRef]
8. Amor, M.D.; Walsh, N.G.; James, E.A. Genetic patterns and climate modelling reveal challenges for conserving *Sclerolaena napiformis* (Amaranthaceae s.l.) an endemic chenopod of Southeast Australia. *Diversity* **2020**, *12*, 417. [CrossRef]
9. Thavornkanlapachai, R.; Levy, E.; Li, Y.; Cooper, S.J.B.; Byrne, M.; Ottewell, K. Disentangling the genetic relationships of three closely related bandicoot species across Southern and Western Australia. *Diversity* **2021**, *13*, 2. [CrossRef]
10. Omland, K.E.; Baker, J.M.; Peters, J.L. Genetic signatures of intermediate divergence: Population history of old and new World holarctic ravens (*Corvus corax*). *Mol. Ecol.* **2006**, *15*, 795–808. [CrossRef] [PubMed]
11. White, D.J.; Ottewell, K.; Spencer, P.; Smith, M.; Short, J.; Sims, C.; Mitchell, N.J. Genetic consequences of multiple translocations of the banded hare-wallaby in Western Australia. *Diversity* **2020**, *12*, 448. [CrossRef]
12. Zilko, J.P.; Harley, D.; Pavlova, A.; Sunnucks, P. Applying population viability analysis to inform genetic rescue that preserves locally unique genetic variation in a critically endangered mammal. *Diversity* **2021**, *13*, 382. [CrossRef]
13. Escalante-Pliego, P.; Matías-Ferrer, N.; Rosas-Escobar, P.; Lara-Martínez, G.; Sepúlveda-González, K.; Raigoza-Figueras, R. Genetic diversity and population structure of Mesoamerican scarlet macaws in an ex-situ breeding population in Mexico. *Diversity* **2022**, *14*, 54. [CrossRef]
14. Glassock, G.L.; Grueber, C.E.; Belov, K.; Hogg, C.J. Reducing the extinction risk of populations threatened by infectious diseases. *Diversity* **2021**, *13*, 63. [CrossRef]
15. Palmas, F.; Righi, T.; Musu, A.; Frongia, C.; Podda, C.; Serra, M.; Splendiani, A.; Barucchi, V.C.; Sabatini, A. Pug-headedness anomaly in a wild and isolated population of native Mediterranean trout *Salmo trutta* L., 1758 complex (Osteichthyes: Salmonidae). *Diversity* **2020**, *12*, 353. [CrossRef]
16. Allendorf, F.W.; Funk, W.C.; Aitken, S.N.; Byrne, M.; Luikart, G. *Conservation and the Genomics of Populations*, 3rd ed.; Oxford University Press: Oxford, UK, 2022.

 *diversity*

Article

Genetic Differentiation among Subspecies of *Banksia nivea* (Proteaceae) Associated with Expansion and Habitat Specialization

Jane Sampson and Margaret Byrne *

Biodiversity and Conservation Science, Department of Biodiversity, Conservation and Attractions, Locked Bag 104, Bentley Delivery Centre, Perth, WA 6983, Australia; jane.sampson@dbca.wa.gov.au
* Correspondence: margaret.byrne@dbca.wa.gov.au

Abstract: Subspecies are traditionally defined using phenotypic differences associated with different geographical areas. Yet patterns of morphological and genetic variation may not coincide and thereby fail to reflect species' evolutionary history. The division of the shrub *Banksia nivea* Labill. into one widespread (*B. nivea* subsp. *nivea*) and two geographically localized subspecies (*B. nivea* subsp. *uliginosa* (A.S. George) A.R. Mast & K.R. Thiele and *B. nivea* subsp. Morangup (M. Pieroni 94/2)) in south-west Australia has been based mainly on variation in leaf shape and pistil length, although flowering time and habitat differences are also evident, and subsp. *uliginosa* occurs on a different substrate. To assess the genetic divergence of *B. nivea* subspecies, we genotyped representatives from each subspecies for nuclear microsatellite and non-coding chloroplast sequence variation. We used distance and parsimony-based methods to assess genetic relatedness. Patterns were consistent with the existing taxonomy of subsp. *nivea* and *uliginosa* but not subsp. Morangup. Phylogenetic analyses revealed evidence for a more recent divergence of subsp. *uliginosa* associated with expansion from dryer sandy soils into the winter-wet ironstone soils in the southwest of Western Australia, consistent with progressive long-term climatic drying. Nuclear microsatellites showed low to moderate diversity, high population differentiation overall, and genetic structuring of subspecies in different biogeographical areas. We propose this pattern reflects the predicted impact of a patchy distribution, small populations, and restrictions to gene flow driven by both distance and biogeographic differences in subspecies' habitats.

Keywords: climate drying; cpDNA; ecotype; Evolutionarily Significant Units; gene flow; geographic expansion; patchy abundance; phylogeography

Citation: Sampson, J.; Byrne, M. Genetic Differentiation among Subspecies of *Banksia nivea* (Proteaceae) Associated with Expansion and Habitat Specialization. *Diversity* **2022**, *14*, 98. https://doi.org/10.3390/d14020098

Academic Editor: Hong-Hu Meng

Received: 19 December 2021
Accepted: 27 January 2022
Published: 30 January 2022

Publisher's Note: MDPI stays neutral with regard to jurisdictional claims in published maps and institutional affiliations.

1. Introduction

Many recognized species are not genetically uniform and may be highly structured into historically isolated populations that may warrant consideration as intraspecific units [1]. Taxonomically recognized subspecies are often based on geographically discontinuous morphological differences [2,3] or ecotypic differences [4]. Yet natural phenotypic or ecotypic diversity within species over wide distributions may not be consistent with genetic divergence representing the evolutionary processes within species [5,6]. Genetic divergence within species is influenced by gene flow [7] and affected by geographic (e.g., topography, distance) and environmental (habitat, climate, pollen, and seed dispersal) factors [8]. Although reciprocal monophyly is not expected for subspecies, some evidence of restricted gene flow between diverging taxa is expected in patterns of neutral genetic variation [6]. Given there are examples of lineage divergence associated with habitat specialization [9–12], widespread species containing subspecies that occupy different habitats might be expected to show genetic differentiation among habitats.

In Australian plant species, distinct population groups and divergent lineages have been identified within species of several genera that reflect disjunct and historically isolated population systems, geological and edaphic complexities, and contrasting habitats in terms of vegetation and climate [13]. Genetically distinct populations are particularly evident within species from the South-West Australia Floristic Region (SWAFR) [14,15] in a range of plant genera and families, including several in genera in Proteaceae, e.g., *Banksia* [12,16] and *Hakea* [17]. Taxonomic resolution can be challenging in this region as phylogenetic lineages often differ from phenotypic variation [17], and the prevalence of highly structured populations and divergent lineages highlights a need to recognize organized layers of genetic diversity below the species level [1].

Banksia nivea Labill. is a common and widespread non-lignotuberous, woody, evergreen, flowering shrub endemic to south-western Australia. There is considerable morphological variation in *B. nivea* mainly in pistil length, leaf size, and shape, and three subspecies are recognized with differing geographic extent [18]. The more common *Banksia nivea* Labill. subsp *nivea* is widespread but patchily distributed in dry sandy soil on sandplains, forests, and mallee from Eneabba in the north, southeast to Cape Arid. Subspecies Morangup (M. Pieroni 94/2) is an informally named subspecies, that is hypothesized to be a unique subspecies but has not been formally assessed or described. It has a highly restricted distribution as it is found at one location in the center of subsp. *nivea*'s range. The rare subspecies *uliginosa* (A.S. George) A.R. Mast & K.R. Thiele is also found in patchy populations in shrublands, woodlands, and forests, and is isolated from subsp. *nivea* and Morangup by >100 km. This subspecies has a limited distribution within a relatively rare, specific edaphic habitat of winter-wet ironstone in two areas on the coastal plain around Busselton and the Scott River in the southwest corner of Western Australia. It has a conservation status of endangered. A dated molecular phylogeny [19] suggests divergence of subsp. *nivea* and *uliginosa* during the Pleistocene (<2.5 Mya).

Unusually for Western Australia flora, *B. nivea* has adaptation to promote gene flow through seed dispersal. Infructescences of *B. nivea* are serotinous containing seeds with a delicate papery wing [20]. Although some follicles open to release seeds periodically, many remain on the plant for a long period of time. If plants die after a disturbance, mass recruitment occurs from the store of genetically diverse seeds held on the plant. Most recruitment occurs after a fire, and the removal of the foliage by fire enables effective wind dispersal of the released seeds [20].

The ecological characteristics of *B. nivea* have been well described [18,21], yet less is known of the species' reproductive biology. Flowers are yellowish-brown in subsp. *uliginosa* and cream-yellow-orange-pink/red-brown in subsp. *nivea* and subsp. Morangup, with a maroon-colored style and green pollen presenter; they have a mousey odor, and are well hidden within the plant [18,22,23]. Recent studies of subsp. *uliginosa* found that plants are self-fertile, and the mating system is predominantly outcrossed with pollination primarily by small non-flying mammals, e.g., honey possums, with some contribution from birds and insects [20,23]. Subspecies *nivea* and *uliginosa* flower between July and September and, although records are few, subspecies Morangup has been recorded to flower in April, June, August, and September [24]. Current flowering periods reported for all subspecies overlap (Western Australian Herbarium) suggesting that any gene flow between subsp. *uliginosa* and the other subspecies is more likely to be restricted by distance and environment rather than temporal isolation. A review of the Australian flora [25] found abundance and population disjunction to be strong determinants of the distribution of contemporary genetic diversity. Given the patchy abundance of *B. nivea* subspecies and the allopatric distributions of ssp. *nivea* and *uliginosa* in different biogeographic areas, we predict that there will be a significant genetic divergence between populations and subspecies. Analysis of the general pattern of genetic variation would provide a phylogenetic context for understanding subspecies relationships and their morphological and ecotypic variation, assist in consideration of the status of subspecies Morangup and provide information for the conservation of the rare subspecies.

Here, we surveyed four cpDNA sequences and 10 nuclear microsatellite loci from multiple individuals of each of the three subspecies of *B. nivea* in populations representative of the species' range to describe the pattern of genetic variation and assess whether it is reflected in the current taxonomy based on the phenotypic and ecotypic variation. Specifically, we address the following questions: (1) is there evidence of genetic structure in the range of *B. nivea*, and (2) are the phenotypically and ecotypic based subspecies reflected in the genetic relationships among populations within the species?

2. Materials and Methods

2.1. Sampling and Genotyping

Leaves were sampled from four to 20 adult plants in seven (subsp. *nivea*), one (subsp. Morangup), and 10 (subsp. *uliginosa*) populations from the ranges of the subspecies (Figure 1). Genomic DNA was extracted from lysed, freeze-dried leaf material following the CTAB-PVP method [26]. The chloroplast *psbA-trnH*, ndhF, trnV, and *trnQ-rps*16 intergenic spacer regions were selected for amplification and sequencing in three random samples from all study populations. Sequence amplification and analysis were conducted according to [27] and sequenced via Macrogen Inc. (Seoul, Korea). SEQUENCHER 5.0 (Genecodes Corp., Ann Arbor, MI, USA) was used to edit miscalls and to align and trim sequences. All four cpDNA regions were concatenated in MESQUITE 3.04 [28] to a total sequence length of 2457 bp, and two indels were identified and coded. Chloroplast haplotypes were identified using DNASP 5.1.1 [29]).

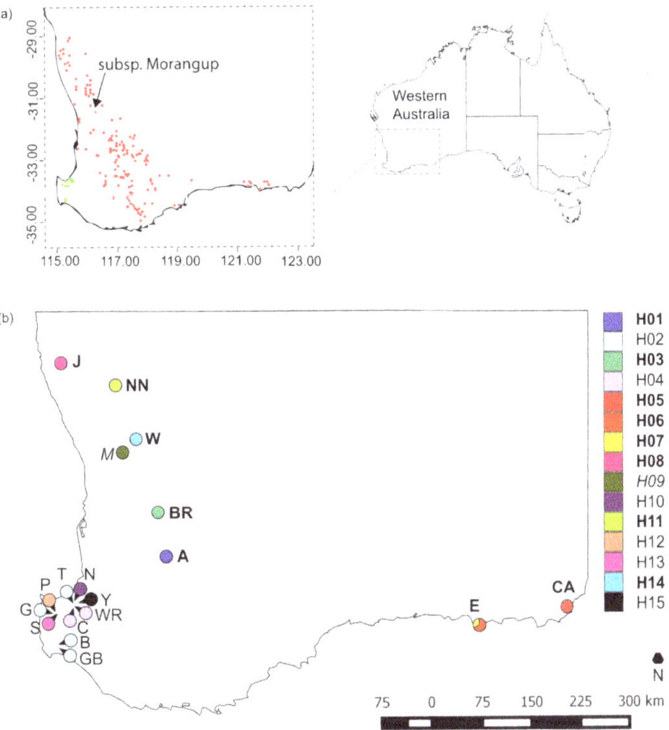

Figure 1. Map showing: (**a**) the distribution of *Banksia nivea* in southwestern Australia. Dots indicate the locations of known populations based on records of the Atlas of Living Australia [30]. Subspecies indicated by color; red, subsp. *nivea*; green, subsp. *uliginosa*; blue, subsp. Morangup. (**b**) Distribution of cpDNA (*psbA-trnH*, ndhF, trnV, and *trnQ-rps*16) haplotypes overlaid on a geographical map of sampling sites. Pie charts show the proportion of sampled individuals with a given haplotype.

Genotypes at 10 nuclear microsatellites were determined in seven subsp. *nivea*, one subsp. Morangup, and seven subsp. *uliginosa* populations (Table 1). Microsatellite loci were amplified as previously described [23,31]. One additional microsatellite locus was also amplified: DnB003; forward primer sequence 5′-AAGCCCAATATGACCAATAACC-3′ and reverse primer sequence 5′-GTCGGCTATATGACTGCATCAC-3′. Modifications to the cited methods were made for $MgCl_2$ concentration and adjusted to 1.5 mM for DnC010 and 1.0 mM for DnB003.

Table 1. Diversity statistics based on 10 nuclear microsatellite loci for populations of *Banksia nivea* from south-western Australia.

Subspecies/Pop	N	A	A_R	H_O	UH_e	F
Subsp. *nivea*						
J	19.900 (0.100)	3.900 (0.994)	2.080 (0.273)	0.351 (0.081)	0.374 (0.083)	0.036 (0.045)
NN	19.600 (0.400)	5.800 (0.663)	3.070 (0.294)	0.498 (0.085)	0.627 (0.081)	0.185 (0.087) *
W	18.100 (0.737)	6.700 (0.831)	3.370 (0.273)	0.566 (0.070)	0.697 (0.058)	0.173 (0.072) *
BR	19.100 (0.605)	6.700 (1.146)	3.120 (0.344)	0.435 (0.085)	0.625 (0.068)	0.248 (0.118) *
A	17.900 (0.623)	8.300 (1.126)	3.640 (0.340)	0.462 (0.073)	0.714 (0.069)	0.281 (0.105) *
E	3.300 (0.578)	2.300 (0.496)	12.630 (0.285)	0.363 (0.098)	0.409 (0.103)	−0.046 (0.098)
CA	10.000 (0.000)	3.500 (0.734)	2.300 (0.367)	0.350 (0.107)	0.404 (0.102)	0.101 (0.098)
Mean	15.414 (0.107)	5.314 (0.092)	4.316 (0.015)	0.432 (0.005)	0.550 (0.007)	0.140 (0.009) *
Subsp. Morangup						
M	19.100 (0.605)	5.300 (0.932)	2.840 (0.321)	0.433 (0.089)	0.574 (0.083)	0.185 (0.117)
Subsp. *uliginosa*						
N	18.900 (0.100)	3.800 (0.533)	2.210 (0.215)	0.401 (0.064)	0.441 (0.073)	0.018 (0.084)
T	17.600 (0.933)	4.100 (0.640)	2.620 (0.283)	0.510 (0.071)	0.548 (0.074)	0.009 (0.070)
Y	10.400 (0.306)	3.000 (0.365)	2.080 (0.267)	0.335 (0.072)	0.366 (0.086)	−0.036 (0.076)
C	18.000 (1.022)	4.700 (0.667)	2.810 (0.239)	0.472 (0.066)	0.593 (0.058)	0.164 (0.097)
G	14.800 (2.489)	3.400 (0.653)	8.330 (0.245)	0.452 (0.077)	0.498 (0.086)	0.029 (0.047)
GB	11.400 (0.909)	4.100 (0.605)	2.650 (0.279)	0.448 (0.065)	0.553 (0.074)	0.107 (0.071)
B	15.800 (1.245)	4.300 (0.517)	2.740 (0.264)	0.404 (0.051)	0.582 (0.067)	0.195 (0.116)
Mean	15.271 (0.292)	3.914 (0.040)	3.349 (0.009)	0.432 (0.003)	0.511 (0.004)	0.069 (0.008)
Total mean	15.593 (0.441)	4.660 (0.227)	3.766 (0.744)	0.432 (0.020)	0.534 (0.021)	0.117 (0.024)

N, mean sample size per locus; A, mean number of alleles per locus; A_R allele number adjusted by rarefaction; H_O, observed heterozygosity; UH_e, unbiased expected heterozygosity; F, Wright's Inbreeding coefficient; standard errors in parentheses; * Significantly different from zero, $p < 0.05$.

Microsatellite loci were separated on an Applied Biosystems (Foster City, CA, USA) 3730 capillary sequencer, and 260 individuals (115 subsp. *nivea*, 20 subsp. Morangup, and 125 subsp. *uliginosa*) were genotyped using GENEMAPPER version 5.0 (Applied Biosystems, Foster City, CA, USA). Tests for stutter bands, large allele dropout, and null alleles were conducted using MICROCHECKER 2.2.3 [32]. Tests of linkage disequilibrium among pairs of loci were performed with GENEPOP 4.2 [33].

2.2. Chloroplast DNA Diversity and Relatedness

We used ARLEQUIN 3.5.2.2 [34] to estimate cpDNA genetic diversity as nucleotide diversity (π), haplotype diversity (H_D), population differentiation as F_{ST}, and differentiation between subspecies (global and pairwise) as F_{ST} using pooled data. Partitioning of diversity between subspecies, populations within subspecies, and within populations was examined by analysis of molecular variance (AMOVA). Estimates of population differentiation (G_{ST}, N_{ST}) and presence of phylogeographical structure ($N_{ST} > G_{ST}$) were estimated using PerMUT 2.0 [35]. If N_{ST} is significantly greater than G_{ST}, haplotypes within populations are more likely to be closely related than haplotypes among populations.

In ARLEQUIN, tests for neutrality and population expansion were estimated using Tajima's D [36] and Fu's F_s [37], and mismatch distribution analyses were also made to infer spatial and demographic history. Goodness-of-fit to models of spatial or demographic expansion were tested with Harpending's raggedness index (H_{Rag}) and the sum of squared differences (SSD). These models test the deviation of observed values from distributions expected under population expansion.

To examine the evolutionary relatedness of chloroplast haplotypes, we constructed a median-joining maximum parsimony (MJMP) network in NETWORK 5.0 [38].

2.3. Nuclear SSR Diversity and Structure

We measured nSSR genetic variation for each subspecies as mean multi-locus parameters per population (number of alleles per locus, A; observed heterozygosity, H_o, unbiased expected heterozygosity, UH_e; Wright's inbreeding coefficient, F) using GENALEX v6.501 [39], and as allele number adjusted by rarefaction AR using HP-RARE [40]. We compared parameters between species using ANOVA.

We measured overall differentiation among populations (F_{ST}) using FREENA with and without the Excluding Null Alleles method (ENA) that corrects for null alleles, with 1000 bootstraps to generate 95% confidence intervals [41]. Partitioning of diversity between subspecies, populations within subspecies, and within populations was examined by analysis of molecular variance (AMOVA) in GENALEX v6.501 [39]. We also estimated global and pairwise differentiation between populations and between subspecies from pooled data as F_{ST} using GENALEX with statistical testing by random permutations. We used a Mantel procedure to test for a correlation between $\log_{10}$ pairwise geographic distances and linearized pairwise genetic distances estimated in GENALEX.

Genetic structure was examined using both direct phenetic and model-based Bayesian analyses. For phenetic analysis, we used PHYLIP 3.69 [42] to construct an unrooted neighbor-joining (NJ) tree based on CS Chord genetic distance calculated in MSA 4.05 [43] with clustering patterns validated with 1000 bootstraps. For Bayesian analyses, we used STRUCTURE 2.3.4 [44] to identify genetically similar clusters (K) and the proportions of individuals' genotypes belonging to clusters (q). To identify the optimum number of clusters (K) and the likelihood of sub-clusters, we used both the ΔK statistic of [45] and the median of estimated Ln probabilities of K values using CLUMPAK [46]. Two or more optimal K may be found if samples are taken from hierarchically structured groups such as species containing subspecies ([45,47]. We, therefore, used hierarchical analyses for (1) all samples and (2) two identified population groups (subsp. *nivea* together with subsp. Morangup, and subsp. *uliginosa* separate). We ran 20 replicates with a burn-in of 100,000 with 500,000 iterations for Markov chain Monte Carlo parameters for $K = 1–15$ possible clusters. We used parameter recommendations [47] including no prior knowledge, the alternative ancestry before separate alphas for each population, an initial ALPHA value of 0.1, and the correlated allele frequency models.

3. Results

3.1. Chloroplast Diversity and Divergence

The concatenated, aligned cpDNA sequence data (2457 bp) revealed 15 haplotypes from 54 individuals from 18 populations: eight haplotypes from subsp. *nivea*, one from subsp. Morangup, and six from subsp. *uliginosa* (Figure 1). Only one population had more than one haplotype; a subsp. *nivea* population (E) was found near the eastern margin of the species' distribution (Figure 1b) with two H06 samples, and one H07 sample. No haplotypes were found in more than one subspecies (Figure 1b) and there was no haplotype diversity in subsp. Morangup. Within the population, haplotype diversity was very low, and most haplotypes were population-specific (86.7%) although two haplotypes were found in multiple subsp. *uliginosa* populations (Figure 1b). Measures of haplotype diversity were therefore high overall, and high in subsp. *nivea*, although lower, although not significantly so, in subsp. *uliginosa* (Table 2). Nucleotide diversity was very low overall and within subspecies. Estimates of overall population differentiation were very high ($F_{ST} = 0.992$, $G_{ST} = 0.961$; Table 2). Tests revealed significant phylogenetic structure in subsp. *nivea* ($N_{ST} > G_{ST}$) but not in subsp. *uliginosa* or overall. AMOVA indicated there was significant differentiation among subspecies (23.42%) although most variation was between populations (75.82%) with a very small proportion (0.76%) within populations (Table 3). Pooled pairwise differentiation was highest for subsp. *uliginosa* compared to subsp. Morangup (F_{ST} = 0.678). Differentiation between subsps. *nivea* and Morangup (F_{ST} = 0.225), and between subsps. *nivea* and *uliginosa* (F_{ST} = 0.270) were lower and similar. This pattern reflects the prevalence of population-specific haplotypes and single-haplotype populations.

Table 2. Diversity statistics, tests of neutrality, and mismatch analyses based on sequences *psbA-trn*H, ndhF, trnV, and *trn*Q-*rps*16 of chloroplast intergenic spacers regions in *Banksia nivea*.

	Total	Subsp. *nivea*	Subsp. *uliginosa*	Subsp. Morangup
Populations (*n*)	18	7	10	1
Haplotypes (*n*)	15	8	6	1
Haplotype diversity H_D	0.920 (0.020)	0.910 (0.026)	0.786 (0.053)	0
Nucleotide diversity π	0.003 (0.002)	0.005 (0.003)	0.001 (0.001)	0
Population differentiation (unordered) G_{ST}	0.961 (0.039)	0.905 (0.095)	1	-
Population differentiation (ordered) N_{ST}	0.992 (0.008)	0.982 (0.018)	1	-
Phylogenetic structure ($N_{ST} > G_{ST}$)	NS	$p < 0.01$	NS	-
Tajima's D	$-0.56\ p = 0.33$	$0.16\ p = 0.61$	$0.14\ p = 0.59$	-
Fu's F_s	$1.13\ p = 0.70$	$4.67\ p = 0.98$	$1.62\ p = 0.77$	-
Demographic expansion	SSD 0.010, $p = 0.61$ H_{Rag} 0.021, $p = 0.43$	SSD 0.034, $p = 0.12$ H_{Rag} 0.045, $p = 0.17$	SSD 0.257, $p = 0.38$ H_{Rag} 0.093, $p = 0.25$	-
Spatial expansion	SSD 0.009, $p = 0.74$ H_{Rag} 0.021, $p = 0.56$	SSD 0.032, $p = 0.20$ HRag 0.045, $p = 0.26$	SSD 0.022, $p = 0.50$ H_{Rag} 0.093, $p = 0.52$	-

Standard deviations are in parentheses. SSD = sum of squared differences. H_{Rag} = Harpending's raggedness index. NS = not significant.

Table 3. Analysis of molecular variance (AMOVA) of *Banksia nivea* subspecies based on chloroplast haplotypes and nuclear microsatellite loci.

Source of Variation	d.f.	SS	Variance Component	% Variation
Chloroplast haplotypes				
Among subspecies	2	27.6	0.569	23.42
Among populations within subspecies	15	83.2	1.842	75.82
Within populations	36	0.7	0.019	0.76
Nuclear microsatellites				
Among subspecies	2	127.3	0.126	3.0
Among populations within subspecies	12	481.5	1.100	26.7
Within populations	505	1463.2	2.897	70.3

The haplotype network was an asymmetrical star structure and contained two closed loops indicative of homoplasy (Figure 2). One side of the network corresponds to subsps. *nivea* and Morangup and showed longer branches with more divergent haplotypes connected through the H01 haplotype to weakly diverged haplotypes found in subsp. *uliginosa*. The haplotype found in subsp. Morangup was divergent but no more so than other haplotypes from subsp. *nivea*. Only H01 was inferred to give rise to several other haplotypes. Therefore, there was a weak geographical pattern concordant with the taxonomy of the described subspecies with haplotypes from subsp. *uliginosa* found in the southwest of the species' range is closely related to haplotypes from subsp. *nivea* from the central part of the species range.

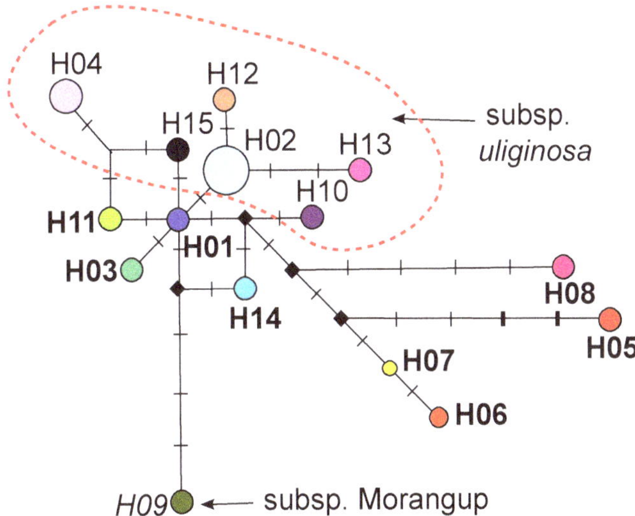

Figure 2. Median-joining maximum parsimony (MJMP) network of chloroplast haplotypes observed in *Banksia nivea* from south-western Australia. Circle sizes are proportional to haplotype frequency among samples. Black dashes represent a single nucleotide substitution (not bold) or indel (bold). Black boxes represent inferred unsampled haplotypes. Haplotype numbers and colors correspond to those in Figure 1.

Estimates of neutrality (Tajima's D, Fu's F_s) that can predict population size changes were not significant for either the species overall or for the separate subspecies (Table 2). However, mismatch analysis revealed evidence of demographic and spatial expansion, both overall and for subsp. *nivea* and *uliginosa* separately, as SSD and H_{RAG} values that did not deviate from models of sudden expansion (Table 2). Non-significant *p*-values do not deviate from the null model and therefore support an expansion scenario.

3.2. Nuclear Diversity and Structure

No evidence of stutter or large allele dropout was detected for nSSR loci and there was no evidence of significant linkage disequilibrium. We detected significant frequencies of null alleles at four loci (DnA011, DnC010, DnB003, and DnD007) but a comparison of F_{ST} (95% CI) estimates with and without ENA adjustment (reported here) showed that null alleles did not cause significant bias and therefore loci were not excluded from analyses.

The average number of alleles per population was low (2.3–8.3; Table 1) even after rarefaction (2.3–12.63). Other diversity measures were moderate and although levels were generally lower in subsp. *uliginosa* than in subsp. *nivea*, there were no significant differences between subspecies. Significant inbreeding was not detected in subsp. *uliginosa* or Morangup although Wright's inbreeding coefficients (*F*) were positive and significant in four of seven populations of subsp. *nivea* (NN, W, BR, A; Table 1).

Genetic differentiation was high overall (F_{ST} = 0.306). Differentiation levels between populations within subspecies were similar (ssp. *nivea* F_{ST} = 0.254; ssp. *uliginosa* F_{ST} = 0.295). Partitioning by AMOVA found only a small amount of variation between subspecies (3%; Table 3). Most variation was within populations (70.3%) with 26.7% between populations. This is reflected in low pooled global differentiation among subspecies (F_{ST} = 0.095). Pairwise pooled comparisons showed the highest differentiation between subsp. Morangup and *uliginosa* (F_{ST} = 0.178), while differentiation between subsp. *nivea* and both Morangup (F_{ST} = 0.094) and *uliginosa* (F_{ST} = 0.082) was lower and similar.

The phenetic analysis showed separation of subsp. *nivea* and subsp. *uliginosa* populations that, although weak, was concordant with taxonomy while subsp. Morangup was nested within subsp. *nivea* (Figure 3a). Grouping was not strong except for the most northern (J, NN) and most south-western (B, GB) populations. There was a significant signal of increasing genetic distance with geographic distance (IBD) across all populations ($r^2 = 0.128$, $p < 0.05$) and also within subspecies (subsp. *nivea*, $r^2 = 0.296$, $p < 0.01$; subsp. *uliginosa*, $r^2 = 0.308$, $p < 0.05$).

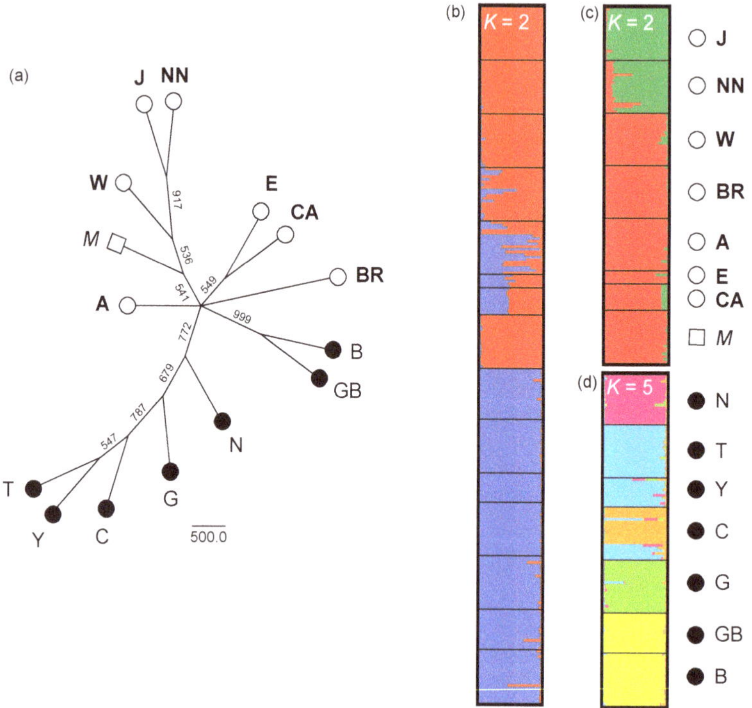

Figure 3. The genetic structure of sampled populations of *Banksia nivea* in south-western Australia, based on individual nuclear microsatellite genotypes. (**a**) A neighbor-joining (NJ) tree of CS Chord distance. Support is shown on the branches as the number of bootstraps out of 1000. Values > 500 are shown. (**b–d**) Bar graphs inferred using Bayesian assignment in STRUCTURE 2.3.4 showing (**b**) structure at $K = 2$ for all subspecies (**c**) structure at $K = 2$ for subspecies *nivea* and Morangup, and (**d**) structure at $K = 5$ for subspecies *uliginosa*. Each individual is represented as a single line with colored segments representing the proportion of ancestry from k clusters (q). Results are the optimal alignment of replicates.

At the highest level, STRUCTURE analysis identified an optimum of two genetic clusters (Figure 3b) that reflected a geographic pattern and were partly aligned with the morphological subsp. *nivea* and *uliginosa*. As in the distance-based analysis, subsp. Morangup clustered with subsp. *nivea*. Admixture was evident in populations in the geographically central (A and BR) and eastern (CA, E) range while populations in the far north and south-west were more clearly differentiated. This may reflect IBD in the center of the range. At a lower level of structure, clusters reflected populations or groups of geographically proximal populations (Figure 3c).

4. Discussion

Patterns of genetic variation between subsp. *nivea* and *uliginosa* were concordant with the current taxonomy, while subsp. Morangup could not be distinguished from subsp. *nivea* in this study. Phylogeographic analyses suggested divergence of subspecies may have been associated with expansion into the southwest corner of Western Australia, a biogeographic area characterized by different substrates, climate, and vegetation. As predicted, patchy abundance was associated with high differentiation between populations and low to moderate nuclear variation reflecting the impact of small population size and restrictions to gene flow. Although genetic diversity was generally lower in the localized subsp. *uliginosa* than in the widespread subsp. *nivea*, differences associated with range were not significant, contrary to expectations based on meta-analysis across Australian plants [25]. This may be due to the stronger effects of patchy distribution in both subspecies.

4.1. Distinction between Subspecies

The morphological differentiation of allopatric subsp. *nivea* and *uliginosa* were reflected in patterns of cpDNA and nuclear microsatellite variation, while subsp. Morangup was not genetically differentiated from subsp. *nivea*. Evolutionary forces act on phenotypic traits differently from neutral genetic markers, and thus morphological divergence is not always associated with genomic divergence. Subspecies have traditionally been defined as phenotypically distinct, allopatric sets of populations that may intergrade at geographic boundaries [2,3] and are widely adopted in plant taxonomy, primarily using geographical and ecological differences to distinguish them [4]. Given the lack of genetic differentiation, we suggest a review of the morphological variation in subsp. Morangup compared to subsp. *nivea* is required to inform a taxonomic revision. A review of approaches to dealing with species-population continuums of genetic diversity for conservation in the age of genomics [1] proposed that, although they may not represent historically isolated populations and satisfy criteria for Evolutionarily Significant Units (ESUs), the recognition of phenotypically defined subspecies may be warranted.

Although differentiated by patterns in the haplotype network and structure plots of nuclear variation, the divergence between subsp. *nivea* and *uliginosa* was not strong. Investigations into congruence between genetic structure and morphological variation in widespread plant species in the SWAFR vary. For example, studies of two widespread and morphologically variable species complexes, *Melaleuca uncinata* R.Br. [48] and *Calothamnus quadrifidus* R.Br. [49] have found general agreement between morphological variation and genetic structure. In contrast, deep lineage divergence in *B. sessilis* [12] was not aligned with morphological varieties and was associated with differences in habitat and substrate.

4.2. Biogeographic Expansion

Differentiation of subsp. *nivea* and *uliginosa* was consistent with other genetic and phylogeographic studies in Australian plants that have identified distinct lineages that reflect the geological and edaphic complexities of habitats that vary in terms of vegetation and climate [14,17,25]. In *B. nivea*, the patterns of geographic separation in different habitats (dry sandy soils contrasting with winter-wet ironstone soils), combined with haplotype diversity and morphological differences provide support for hypotheses of divergence driven by expansion into a different biogeographic area. Both subsp. *nivea* and *uliginosa* showed high haplotype diversity with low nucleotide diversity, signals of expansion, and a phylogeographic pattern consistent with subspecies differentiation. The phylogeographic structure was also observed in the more widespread subsp. *nivea*. These traits suggest that diversification associated with geographic isolation and habitat specificity are likely to have contributed to the divergence of subspecies following expansion into the distinct southwest Australian ironstone habitat. A similar scenario was proposed [49] for another ironstone endemic, the woody shrub *Calothamnus quadrifidus* subsp. *teretifolius* A.S. George & N. Gibson, which does not share haplotypes with other subspecies of *C. quadrifidus* found outside the ironstone habitat. Similar genetic differentiation of populations occurring in the specific ironstone habitat has also been observed in *Hakea oldfieldii* Benth. [50]. Expansion to occupy habitats with different substrate likely leads to differentiation through adaptation, and divergence associated with substrate has also been identified in other species in different ecosystems, e.g., [9–12].

In *B nivea*, a more ancient origin for subsp. *nivea* compared to subsp. *uliginosa* is suggested by higher diversity and mutational divergence of haplotypes compared to subsp. *uliginosa*. This would be consistent with branching patterns in the dryandra clade of the *Banksia* phylogeny [19]. Although major spatial contraction-expansion dynamics appear to have been rare in the mesic biota of Australia [51,52], range expansions have been associated with acceleration of the progressive drying of mesic environments that began in the late Pliocene (c. 3 Mya) [52,53]. Range expansions have also been associated with the southward progress of increasing aridity in the SWAFR from the mid-Pleistocene that opened up habitats within the wetter forests allowing for the expansion of species to these areas [49,54,55]. Expansion and divergence in *B. nivea* are consistent with this pattern providing another example of expansion among widespread, woody shrubs. Indeed, the pattern of southwestern expansion, as reflected in haplotype network relationships and divergence, is most similar to that found in the widespread wind-pollinated shrub, the dwarf sheoak *Allocasuarina humilis* (Otto & A. Dietr.) L.A.S. Johnson [54], although on a smaller scale. This phylogeographic pattern in *A. humilis* was also best explained by south-west expansion from populations with a more ancient history of persistence in dry shrublands (300–600 mm year^{-1}) into areas previously occupied by higher rainfall forests (600–1400 mm year^{-1}) as the climate dried progressively from the late Miocene [54].

4.3. Contemporary Genetic Diversity and Population Structure

Many reviews of nuclear genetic variation have found that genetic structure is influenced by mating systems, life-history traits, chromosomal variation, population distribution, and other ecological traits related to gene flow [7,56–59]. A specific analysis of diversity and population differentiation in the Australian flora [25] found strong associations with abundance, where patchy populations were significantly associated with low diversity and high differentiation. We found patterns consistent with this in *B. nivea* but, contrary to the strong association also expected for range and diversity, both localized and widespread subspecies of *B. nivea* had diversity levels that are expected for localized species. Over the widespread distribution of the species, low to moderate nuclear genetic diversity and moderate to high differentiation among populations is not unexpected. Patchy populations are often small, and population genetic theory predicts they may be prone to genetic drift and inbreeding leading to loss of genetic diversity and differentiation [56,60]. We found no significant evidence of drift or bottlenecks in *B. nivea*, although our ability to detect these impacts was limited by sample size. We did identify significant inbreeding in this study in some populations of subsp. *nivea* but not in subsp. *uliginosa*. More detailed studies of the mating system in subsp. *uliginosa* showed high outcrossing (95%) and little relatedness amongst adult plants within populations [23], and high production of seeds that was unaffected by population size [20].

Analysis of contemporary genetic relationships revealed evidence of restricted gene flow between populations across the species' range with geographic clustering, high pairwise F_{ST} values, a significant proportion of genetic variation apportioned between populations, and significant IBD. We detected a genetic pattern in cluster analysis and in the distance-based tree that related to the current subspecies taxonomy and different biogeographic areas. This, together with significant IBD, is likely to reflect restricted gene flow across heterogeneous landscapes. Cluster analysis also detected a geographic pattern in genetic sub-structuring within both subsp. *nivea* and *uliginosa* that are likely to reflect gene flow patterns. Gene flow via pollen dispersal is achieved in *Banksia* species by birds, mammals, and insects [61,62]. Specific pollinators for *B. nivea* have not been determined and it is likely pollination is achieved primarily by small non-flying mammals that move within a small range (<30 m), even over several months, and are likely to achieve near-neighbor pollination, and by birds such as honeyeaters that are likely to facilitate some longer distance pollen dispersal between local populations. In a study on pollinators in subsp. *uliginosa* [20] treatments open to birds and mammals produced high levels of fruit compared to those open to invertebrates only or closed to all pollinators, and treatments open to all pollinators produced 39% more fruit than those open to mammals but not birds. Effective pollination was also shown in a study of the mating system that found high levels of outcrossing (95%), and up to 30% of progeny produced in seed crops was attributed to mating with fathers outside small patchy populations likely due to bird pollination [23]. Unusually for *Banksia*, seeds of *B. nivea* have adaptations for dispersal, and the sharing of seed-dispersed haplotypes among geographically close subsp. *uliginosa* populations may reflect some localized inter-population dispersal, although this is difficult to distinguish from co-ancestry among local populations. The significant differentiation among populations and the biogeographic structure observed in *B. nivea* likely reflects generally localized pollen dispersal associated with habitat specificity of the predominant non-flying mammal pollinators [20,23], along with generally short-range gene dispersal by seeds.

5. Conclusions

Analysis of genetic relationships among the three subspecies of *B. nivea* in south-western Australia supported the current taxonomy of subspecies *nivea* and *uliginosa*, and indicated clarification of the morphological traits and heritability in subsp. Morangup is warranted. The climatic history of the SWAFR appears to have had a significant influence on the genetic divergence within *B. nivea*. We found patterns of variation consistent with expansion into a new biogeographical area and onto different substrate followed by divergence into lineages concurrent with the taxonomic subspecies. The pattern of nuclear DNA diversity and differentiation likely reflects the influence of distribution and restricted gene flow between small and patchy populations.

Author Contributions: Conceptualization, M.B.; methodology, M.B. and J.S.; formal analysis, J.S.; writing—original draft preparation, J.S.; writing—review and editing, J.S. and M.B., visualization, J.S. All authors have read and agreed to the published version of the manuscript.

Funding: This research received no external funding.

Data Availability Statement: The data presented in this study are available on request from the corresponding author. The data are not publicly available due to legislative protection for threatened species.

Acknowledgments: We thank Sarah Tapper, Ashney Shah, Shelley McArthur, Bronwyn Macdonald, and Margaret Hankinson for laboratory assistance. We thank Neil Gibson and Colin Yates for their assistance in the field and discussion on the biology of *Banksia nivea* and the habitat specialization of the flora.

Conflicts of Interest: The authors declare no conflict of interest.

References

1. Coates, D.J.; Byrne, M.; Moritz, C. Genetic diversity and conservation units: Dealing with the species-population continuum in the age of genomics. *Front. Ecol. Evol.* **2018**, *6*, 165. [CrossRef]
2. Mayr, E. *Systematics and the Origin of Species*; Columbia University Press: New York, NY, USA, 1942.
3. Grant, V. *Plant Speciation*; Columbia University Press: New York, NY, USA, 1981.
4. Hamilton, C.; Reichard, S. Current practice in the use of subspecies, variety, and forma in the classification of wild plants. *Taxon* **2019**, *41*, 485–498. [CrossRef]
5. Duminil, J.; Kenfack, D.; Viscosi, V.; Grumiau, L.; Hardy, O.J. Testing species delimitation in sympatric species complexes: The case of an African tropical tree, *Carapa* spp. (Meliaceae). *Mol. Phylogenet. Evol.* **2012**, *62*, 275–285. [CrossRef] [PubMed]
6. Christmas, M.J.; Biffin, E.; Lowe, A.J. Measuring genome-wide genetic variation to reassess subspecies classifications in *Dodonaea viscosa* (Sapindaceae). *Aust. J. Bot.* **2018**, *66*, 287–297. [CrossRef]
7. Ellstrand, N.C. Is gene flow the most important evolutionary force in plants? *Am. J. Bot.* **2014**, *101*, 737–753. [CrossRef]
8. Wang, I.J.; Bradburd, G.S. Isolation by environment. *Mol. Ecol.* **2014**, *23*, 5649–5662. [CrossRef]

9.   Zhang, Y.-H.; Wang, I.J.; Comes, H.P.; Peng, H.; Qiu, Y.-X. Contributions of historical and contemporary geographic and environmental factors to phylogeographic structure in a Tertiary relict species, *Emmenopterys henryi* (Rubiaceae). *Sci. Rep.* **2016**, *6*, 24041. [CrossRef]

10. Robins, T.P.; Binks, R.M.; Byrne, M.; Hopper, S.D. Contrasting patterns of population divergence on young and old landscapes in *Banksia seminuda* (Proteaceae), with evidence for recognition of subspecies. *Biol. J. Linn. Soc.* **2021**, *133*, 449–463. [CrossRef]

11. Moore, A.J.; Merges, D.; Kadereit, J.W. The origin of the Serpentine endemic *Minuartia larcifolia* subsp. *ophiolitica* by vicariance and competitive exclusion. *Mol. Ecol.* **2013**, *22*, 2218–2231. [CrossRef]

12. Nistelberger, H.M.; Tapper, S.L.; Coates, D.J.; McArthur, S.L.; Byrne, M. As old as the hills: Pliocene palaeogeographical processes influence patterns of genetic structure in the widespread, common shrub *Banksia sessilis*. *Ecol. Evol.* **2021**, *11*, 1069–1082. [CrossRef]

13. Byrne, M.; Murphy, D.J. The origins and evolutionary history of xerophytic vegetation in Australia. *Aust. J. Bot.* **2020**, *68*, 195–207. [CrossRef]

14. Coates, D.J. Defining conservation units in a rich and fragmented flora: Implications for the management of genetic resources and evolutionary processes in south-west Australian plants. *Aust. J. Bot.* **2000**, *48*, 329–339. [CrossRef]

15. Byrne, M.; Coates, D.; Forest, F.; Hopper, S.; Krauss, S.; Sniderman, J.; Thiele, K. A diverse flora—species and genetic relationships. In *Plant Life on the Sandplains in Southwest Australia, a Global Biodiversity Hotspot*; Lambers, H., Ed.; University of Western Australia Publishing: Crawley, WA, USA, 2014; pp. 81–99.

16. Coates, D.J.; McArthur, S.L.; Byrne, M. Significant genetic diversity loss following pathogen driven population extinction in the rare endemic *Banksia brownii* (Proteaceae). *Biol. Conserv.* **2015**, *192*, 353–360. [CrossRef]

17. Sampson, J.F.; Byrne, M.; Yates, C.J.; Gibson, N.; Thavornkanlapachai, R.; Stankowski, S.; Macdonald, B.; Bennett, I. Contemporary pollen-mediated gene immigration reflects the historical isolation of a rare, animal-pollinated shrub in a fragmented landscape. *Heredity* **2014**, *112*, 172–181. [CrossRef] [PubMed]

18. Cavanagh, A.K.; Pieroni, M. *The Dryandras*; Australian Plants Society (SGAP Victoria) Inc.; Wildflower Society of Western Australia Inc.: Melbourne, Australia, 2006.

19. Cardillo, M.; Pratt, R. Evolution of a hotspot genus: Geographic variation in speciation and extinction rates in *Banksia* (Proteaceae). *BMC Evol. Biol.* **2013**, *13*, 155. [CrossRef] [PubMed]

20. Thavornkanlapachai, R.; Byrne, M.; Yates, C.J.; Ladd, P.G. Degree of fragmentation and population size do not adversely affect reproductive success of a rare shrub species, *Banksia nivea* (Proteaceae), in a naturally fragmented community. *Bot. J. Linn. Soc.* **2019**, *191*, 261–273. [CrossRef]

21. George, A. New taxa and a new infrageneric classification in *Dryandra* R.Br. (Proteaceae: Grevilleoideae). *Nuytsia* **1996**, *10*, 313–408.

22. Brown, A.; Thomson-Dans, C.M.N. *Western Australia's Threatened Flora*; Brown, A., Thomson-Dans, C.M.N., Eds.; Department of Conservation and Land Management: Perth, Australia, 1998.

23. Thavornkanlapachai, R.; Ladd, P.G.; Byrne, M. Population density and size influence pollen dispersal pattern and mating system of the predominantly outcrossed *Banksia nivea* (Proteaceae) in a threatened ecological community. *Biol. J. Linn. Soc.* **2018**, *124*, 492–503. [CrossRef]

24. FloraBase-the Western Australian Flora. Available online: https://florabase.dpaw.wa.gov.au/browse/profile/32205 (accessed on 11 June 2021).

25. Broadhurst, L.; Breed, M.; Lowe, A.; Bragg, J.; Catullo, R.; Coates, D.; Encinas-Viso, F.; Gellie, N.; James, E.; Krauss, S.; et al. Genetic diversity and structure of the Australian flora. *Divers. Distrib.* **2017**, *23*, 41–52. [CrossRef]

26. Byrne, M.; Parrish, T.L.; Moran, G.F. Nuclear RFLP diversity in *Eucalyptus nitens*. *Heredity* **1998**, *81*, 225–233. [CrossRef]

27. Byrne, M.; Hankinson, M. Testing the variability of chloroplast sequences for plant phylogeography. *Mol. Ecol. Resour.* **2012**, *60*, 569–574. [CrossRef]

28. Maddison, W.P.; Maddison, D.R. Mesquite: A Modular System for Evolutionary Analysis. Available online: http://mesquiteproject.org (accessed on 2 October 2017).

29. Librado, P.; Rozas, J. DnaSP v.5: A software for comprehensive analysis of DNA polymorphism data. *Bioinformatics* **2009**, *25*, 1451–1452. [CrossRef]

30. Atlas of Living Australia. Available online: https://bie.ala.org.au/species/https://id.biodiversity.org.au/node/apni/2894027 (accessed on 16 June 2021).

31. Millar, M.A.; Byrne, M. Characterization of polymorphic microsatellite DNA markers in *Banksia nivea*, formerly *Dryandra nivea*. *Mol. Ecol. Resour.* **2008**, *8*, 1393–1394. [CrossRef] [PubMed]

32. van Oosterhout, C.; Hutchinson, W.F.; Wills, D.P.M.; Shipley, P. MICRO-CHECKER: Software for identifying and correcting genotyping errors in microsatellite data. *Mol. Ecol. Notes* **2004**, *4*, 535–538. [CrossRef]

33. Rousset, F. GENEPOP'007: A complete reimplementation of the GENEPOP software for Windows and Linux. *Mol. Ecol. Resour.* **2008**, *8*, 103–106. [CrossRef]

34. Excoffier, L.; Lischer, H. Arlequin Suite Ver 3.5: A new series of programs to perform population genetic analyses under Linux and Windows. *Mol. Ecol. Resour.* **2010**, *10*, 464–467. [CrossRef]

35. Pons, O.; Petit, R.J. Measuring and testing genetic differentiation with ordered versus unordered alleles. *Genetics* **1996**, *144*, 1238–1245. [CrossRef]

36. Tajima, F. Statistical methods for testing the neutral mutation hypothesis by DNA polymorphism. *Genetics* **1989**, *123*, 585–595. [CrossRef]

37. Fu, Y.X. Statistical tests of neutrality of mutations against population growth, hitchhiking and background selection. *Genetics* **1997**, *147*, 915–925. [CrossRef]

38. Bandelt, H.-J.; Forster, P.; Rohl, A. median-joining networks for inferring intraspecific phylogenies. *Mol. Biol. Evol.* **1999**, *16*, 37–48. [CrossRef]

39. Peakall, R.; Smouse, P.E. GenALEx 6.5: Genetic analysis in Excel. Population genetic software for teaching and research-an update. *Bioinformatics* **2012**, *28*, 2537–2539. [CrossRef] [PubMed]

40. Kalinowski, S.J. HP-RARE 1.0: A computer program for performing rarefaction on measures of allelic richness. *Mol. Ecol. Notes* **2005**, *5*, 187–189. [CrossRef]

41. Chapuis, M.P.; Estoup, A. Microsatellite null alleles and estimation of population differentiation. *Mol. Biol. Evol.* **2007**, *24*, 621–631. [CrossRef] [PubMed]

42. Felsenstein, J. PHYLIP—phylogeny inference package (Version 3.2). *Cladistics* **1989**, *5*, 164–166.

43. Dieringer, D.; Schlötterer, C. Microsatellite Analyser: A platform independent analysis tool for large microsatellite data sets. *Mol. Ecol. Notes* **2003**, *3*, 167–169. [CrossRef]

44. Pritchard, J.K.; Stephens, M.; Donnelly, P. Inference of population structure using multilocus genotype data. *Genetics* **2000**, *155*, 945–959. [CrossRef] [PubMed]

45. Evanno, G.; Regnaut, S.; Goudet, J. Detecting the number of clusters of individuals using the software STRUCTURE: A simulation study. *Mol. Ecol.* **2005**, *14*, 2611–2620. [CrossRef] [PubMed]

46. Kopelman, N.M.; Mayzel, J.; Jakobsson, M.; Rosenberg, N.A.; Mayrose, I. Clumpak: A program for identifying clustering modes and packaging population structure inferences across K. *Mol. Ecol. Resour.* **2015**, *15*, 1179–1191. [CrossRef]

47. Wang, J. The computer program Structure for assigning individuals to populations: Easy to use but easier to misuse. *Mol. Ecol. Resour.* **2017**, *17*, 981–990. [CrossRef]

48. Broadhurst, L.; Byrne, M.; Craven, L.; Lepschi, B. Genetic congruence with new species boundaries in the *Melaleuca uncinata* complex (Myrtaceae). *Aust. J. Bot.* **2004**, *52*, 729–737. [CrossRef]

49. Nistelberger, H.; Gibson, N.; Macdonald, B.; Tapper, S.-L.; Byrne, M. Phylogeographic evidence for two mesic refugia in a biodiversity hotspot. *Heredity* **2014**, *113*, 454–463. [CrossRef] [PubMed]

50. Sampson, J.F.; Hankinson, M.; McArthur, S.; Tapper, S.; Langley, M.; Gibson, N.; Yates, C.; Byrne, M. Long-term "islands" in the landscape: Low gene flow, effective population size and genetic divergence in the shrub *Hakea oldfieldii* (Proteaceae). *Bot. J. Linn. Soc.* **2015**, *179*, 319–334. [CrossRef]

51. Byrne, M. Evidence for multiple refugia at different time scales during Pleistocene climatic oscillations in southern Australia inferred from phylogeography. *Quat. Sci. Rev.* **2008**, *27*, 2576–2585. [CrossRef]

52. Byrne, M.; Steane, D.A.; Joseph, L.; Yeates, D.K.; Jordan, G.J.; Crayn, D.; Aplin, K.; Cantrill, D.J.; Cook, L.G.; Crisp, M.D.; et al. Decline of a biome: Evolution, contraction, fragmentation, extinction and invasion of the Australian mesic zone biota. *J. Biogeogr.* **2011**, *38*, 1635–1656. [CrossRef]

53. Byrne, M.; Yeates, D.K.; Joseph, L.; Kearney, M.; Bowler, J.; Williams, M.A.J.; Cooper, S.; Donnellan, S.C.; Keogh, J.S.; Leys, R.; et al. Birth of a biome: Insights into the assembly and maintenance of the Australian arid zone Biota. *Mol. Ecol.* **2008**, *17*, 4398–4417. [CrossRef] [PubMed]

54. Llorens, T.M.; Tapper, S.-L.; Coates, D.J.; McArthur, S.; Hankinson, M.; Byrne, M. Does population distribution matter? Influence of a patchy versus continuous distribution on genetic patterns in a wind-pollinated shrub. *J. Biogeogr.* **2016**, *44*, 361–374. [CrossRef]

55. Sampson, J.; Tapper, S.; Coates, D.; Hankinson, M.; McArthur, S.; Byrne, M. Persistence with episodic range expansion from the early Pleistocene: The distribution of genetic variation in the forest tree *Corymbia calophylla* (Myrtaceae) in south-Western Australia. *Biol. J. Linn. Soc.* **2018**, *123*, 545–560. [CrossRef]

56. Loveless, M.; Hamrick, J. Ecological determinants of genetic structure in plant populations. *Annu. Rev. Ecol. Evol. Syst.* **1984**, *15*, 65–96. [CrossRef]

57. Nybom, H. Comparison of different nuclear DNA markers for estimating intraspecific genetic diversity in plants. *Mol. Ecol.* **2004**, *13*, 1143–1155. [CrossRef]

58. Duminil, J.; Fineschi, S.; Hampe, A.; Jordano, P.; Salvini, D.; Vendramin, G.G.; Petit, R.J.; The, S.; Naturalist, A.; May, N. Can population genetic structure be predicted from life-history traits? *Am. Nat.* **2014**, *169*, 662–672. [CrossRef]

59. Gitzendanner, M.A.; Soltis, P.S. Patterns of genetic variation in rare and widespread plant congeners. *Am. J. Bot.* **2000**, *87*, 783–792. [CrossRef] [PubMed]

60. Ellstrand, N.C.; Elam, D.R. Population genetic consequences of small population size: Implications for plant conservation. *Annu. Rev. Ecol. Evol. Syst.* **1993**, *24*, 217–242. [CrossRef]

61. Ford, H.A.; Paton, D.C.; Forde, N. Birds as pollinators of Australian plants. *N. Z. J. Bot.* **1979**, *17*, 509–519. [CrossRef]

62. Wooller, R.D.; Russell, E.M.; Renfree, M.B.; Towers, P.A. A comparison of seasonal changes in the pollen loads of nectarivorous marsupials [*Tarsipes*] and birds [honeyeaters]. *Wildl. Res.* **1983**, *10*, 311–317. [CrossRef]

Article

Genomic Screening Reveals That the Endangered *Eucalyptus paludicola* (Myrtaceae) Is a Hybrid

Kor-jent van Dijk [1], Michelle Waycott [1,2,*], Joe Quarmby [3,4], Doug Bickerton [3], Andrew H. Thornhill [1,2], Hugh Cross [1,5] and Edward Biffin [2]

1   School of Biological Sciences, University of Adelaide, Adelaide, SA 5005, Australia; korjent.vandijk@adelaide.edu.au (K.-j.v.D.); andrew.thornhill@adelaide.edu.au (A.H.T.); hugh.cross@otago.ac.nz (H.C.)
2   Department for Environment and Water, Botanic Gardens and State Herbarium, State Herbarium of South Australia, Adelaide, SA 5001, Australia; edward.biffin@adelaide.edu.au
3   Department for Environment and Water, Government of South Australia, Adelaide, SA 5001, Australia; jquarmby@tasland.org.au (J.Q.); doug.bickerton@sa.gov.au (D.B.)
4   Tasmanian Land Conservancy, Lower Sandy Bay, TAS 7005, Australia
5   Department of Anatomy, University of Otago, Dunedin 9054, New Zealand
*   Correspondence: michelle.waycott@adelaide.edu.au

Received: 12 November 2020; Accepted: 7 December 2020; Published: 10 December 2020

Abstract: A hybrid origin for a conservation listed taxon will influence its status and management options. Here, we investigate the genetic origins of a nationally endangered listed taxon—*Eucalyptus paludicola*—a tree that is restricted to the Fleurieu Peninsula and Kangaroo Island of South Australia. Since its description in 1995, there have been suggestions that this taxon may potentially be a stable hybrid species. Using a high throughput sequencing approach, we developed a panel of polymorphic loci that were screened across *E. paludicola* and its putative parental species *E. cosmophylla* and *E. ovata*. Bayesian clustering of the genotype data identified separate groups comprising *E. ovata* and *E. cosmophylla* while *E. paludicola* individuals were admixed between these two, consistent with a hybrid origin. Hybrid class assignment tests indicate that the majority of *E. paludicola* individuals (~70%) are F_1 hybrids with a low incidence of backcrossing. Most of the post-F_1 hybrids were associated with revegetation sites suggesting they may be maladapted and rarely reach maturity under natural conditions. These data support the hypothesis that *E. paludicola* is a transient hybrid entity rather than a distinct hybrid species. We briefly discuss the conservation implications of our findings.

Keywords: conservation; natural hybridisation; *Eucalyptus*; high throughput sequencing; NGS; South Australia

1. Introduction

Natural interspecific hybridisation is a common phenomenon in plants and is an important evolutionary process. The outcomes of hybridisation can be diverse, maintaining, reducing or increasing evolutionary divergence among taxa [1]. Hybridisation between well-differentiated species can lead to the origin of new species involving a change in base chromosome number (i.e., allopolyploid hybrid speciation) or without such a change (homoploid hybrid speciation). In the case of the former, a change in ploidy can lead to the spontaneous development of reproductive isolation, while in contrast successful homoploid hybrid species must overcome significant ecological, genetic and demographic obstacles and are thus considered rare [2]. Homoploid hybrid speciation has been suggested in a number of cases, although the majority of these have demonstrated a hybrid origin of taxa but lack conclusive evidence for hybrid speciation [1]. Key additional criteria include evidence of reproductive isolation of hybrid lineages from the parental species and evidence that reproductive isolation arose as a consequence of hybridisation [2].

The genus *Eucalyptus L'Hér.* includes approximately 650 species of mostly trees that are endemic to Australia (with a few exceptions) and dominate the forests and woodlands of that continent. Taxonomists classify eucalypt species, including members of the three genera: *Angophora Cav.*, *Corymbia K.D.Hill & L.A.S.Johnson,* and *Eucalyptus*, into various subspecies, sections and series based on morphology and assumed relatedness (see Nicolle 2019 [3] for the most recent classification). Eucalypts are the worlds' most widely grown plantation hardwoods and there has been considerable interest in genetic improvement through manipulated hybridisation [4]. Natural interspecific hybridisation has also been widely reported within the group (e.g., [5–9]) and the propensity for hybridisation and its outcomes are strongly predicted by the degree of evolutionary divergence among species [5,10–13]. Natural hybridisation between species from different genera or subgenera is not believed to occur (but see [7] for a possible exception) and is relatively uncommon among sections within subgenera [5,13]. There are several well-documented cases of natural hybridisation amongst closely related eucalypts although according to Griffin et al. [5] only 15% of cases, where hybridisation was likely (intrasectional relatives with proximal distributions), resulted in hybrid formation. Known prezygotic barriers in eucalypts include pollen tube arrest, the frequency of which increases with evolutionary distance between parents [14] while species-specific variation in flower size presents a structural barrier to gene flow [15]. Postzygotic barriers may also contribute to reproductive isolation. In a recent study, there was strong evidence for outbreeding depression (negative epistasis) amongst two closely related *Eucalyptus* species [16], which is predicted to increase with evolutionary divergence because of "snowballing" epistatic interactions [17].

Eucalyptus paludicola D.Nicolle was described by Nicolle 1995 [18] and in light of its rarity and exposure to ongoing threatening processes [19] has been listed as endangered under the Environment Protection and Biodiversity Conservation Act of 1999 (Commonwealth of Australia) and the National Parks and Wildlife Act 1972 (State of South Australia). The known distribution includes approximately 34 populations with an estimated 720–750 individuals in total [20]. Its range is limited to the Fleurieu Peninsula and Kangaroo Island of South Australia, where it occurs on seasonal swampy sites, often in highly modified landscapes. *E. paludicola* is commonly coassociated with *Eucalyptus cosmophylla F. Muell.* and *E. ovata Labill.* and is in some respects morphologically and ecologically [18] as well as genetically (based on amplified fragment length polymorphisms [21]; and DART markers [22]) intermediate between these species, suggesting the likelihood of a hybrid origin. The relatively constant morphology of *E. paludicola* throughout its distributional range and evidence that the progeny breed true in an arboretum has been argued as supporting specific recognition [18,23]. While existing molecular evidence is consistent with a hybrid origin, a hypothesis of hybrid speciation is difficult to distinguish from alternatives [2], for instance, that *E. paludicola* individuals represent transient hybrids lacking reproductive isolation from the parents. Given that there are presently no known instances of hybrid speciation among eucalypts, a resolution of this issue may have significant bearing on our understanding of hybridisation and speciation within this ecologically and economically important lineage. Resolving the origins of *E. paludicola* is also important in clarifying its taxonomic and conservation status. In Australia, if a species is listed under the Federal Environment Protection and Biodiversity Conservation Act 1999 (EPBC Act), then it must be protected. Further, species that have arisen via hybridisation may be afforded full conservation protection while transient hybrid (i.e., lacking reproductive isolation from the parents and hybridisation not directly leading to the formation of a new taxon) entities are not presently recognised under relevant Commonwealth and State legislation. The EPBC Act does not, however, require a test to identify if that species might be a transient hybrid or hybrid species, meaning some currently listed taxa may be invalid.

To resolve the origins of *E. paludicola* we used a high throughput sequencing (HTS) approach to assess genetic diversity. Specifically, we leveraged the *E. grandis* genome sequence [24] and restriction-site based HTS to develop a panel of molecular markers. We used multiplex PCRs to screen across representatives of *E. paludicola* along with *E. ovata* and *E. cosmophylla* and applied Bayesian statistical approaches to hybrid assessment from these data. If *E. paludicola* is a hybrid species it will be indicated by factors including unique genetic diversity and coherence across its geographical range. The alternative hypothesis, that this taxon is a transient hybrid, would be supported by a predominance of early hybrid generations (e.g., F_1 individuals) consistent with the absence of reproductive isolation.

2. Materials and Methods

2.1. Study Species

We sampled individuals of *E. paludicola*, *E. cosmophylla* and *E. ovata* across the region of geographic sympatry that includes the southern Fleurieu Peninsula and Kangaroo Island in the state of South Australia. Both *E. paludicola* and *E. cosmophylla* are endemic to this region, while *E. ovata* is widely distributed in temperate south-eastern Australia, with a disjunct range extending to our study region. All three species are placed within *Eucalyptus* subg. *Symphyomyrtus*, and *E. paludicola* and *E. cosmophylla* are considered to be close relatives (Section *Incognitae*; [18]) while *E. ovata* is placed within Section *Maidenaria*. *E. paludicola* and *E. cosmophylla* can be readily distinguished by their habit (taller and more upright, versus a shrubby to mallee habit, respectively) and inflorescence structure (usually seven flowered in *E. paludicola* versus consistently three flowered in *E. cosmophylla*), while *E. paludicola* has thinner leaves, longer peduncles and pedicels and smaller flowers and fruits. *E. ovata* has a tree habit, a consistently seven flowered inflorescence and has smaller flowers and fruit than the other two species. Both *E. paludicola* and *E. ovata* are associated with seasonally swampy sites, while *E. cosmophylla* grows on a range of soil types from infertile sands to poorly drained gravelly clays [18,23].

2.2. Sample Collection and DNA Extraction

Leaf material was collected from individuals of each species on both Fleurieu Peninsula and Kangaroo Island and dried using silica-gel. Genomic DNA was subsequently extracted from the leaf material using a commercial service (Australian Genome Research Facility, Adelaide). Sampling details are included in Table 1 (additional details in supplementary Table S1). Our sampling included two sites that were planted with *E. paludicola* as part of a restoration project (*Eucalyptus paludicola* recovery program, Department for Environment and Water).

2.3. Marker Development and Sequencing

We used methods described by Cross et al. [25] to generate a sequencing library for representative samples of *E. cosmophylla*, *E. ovata* and *E. paludicola*. The library was then sequenced using an Ion Torrent Personal Genome Machine (Ion PGM, Life Technologies, Carlsbad, California, USA) with 200 bp sequencing chemistry and a 318 v.1 sequencing chip. The raw sequences were imported into CLC Genomics Workbench 9 (Qiagen, Venlo, The Netherlands) for demultiplexing, adapter and quality trimming (ambiguous trim = 2, quality limit = 0.05, minimum read length = 50). The trimmed reads were then de novo assembled (indel and mismatch cost = 2; similarity fraction = 0.8, length fraction = 0.5) and we extracted contigs with coverage greater than 100×. The resulting contigs were compared to the *E. grandis* v2.0 genome sequence [24] using *blast-n* (default parameters in CLC) to identify regions that, from our sequencing library, had a single blast hit and could be considered as potential single copy regions in our target eucalypt species. From these, we targeted regions that had one or more SNPs in our assemblies and had suitable priming sites (based upon the *E. grandis* genome sequence; supplementary Table S2) to amplify a product of not more than 150 bp (excluding primer, adapter and barcode sequences), which is the approximate limit of the Ion Torrent PGM 200 bp sequencing chemistry. Primers were designed in Geneious v7 (Biomatters, Auckland, New Zealand) [26] using the Primer 3 [27] plug-in.

Table 1. Taxa and locations of tested samples.

Taxon	Location	*n*	Latitude	Longitude
E. cosmophylla	Three Chain Rd (KI)	6	−35.829	137.704
E. cosmophylla	Crafers West (FL)	2	−34.990	138.681
E. cosmophylla	Kyeema (FL)	4	−35.270	138.674
E. cosmophylla	Mt Billy Cp (FL)	2	−35.448	138.600
E. cosmophylla	Burnfoot (FL)	4	−35.400	138.557
E. ovata	Burnfoot (FL)	3	−35.400	138.557
E. ovata	Cleland Gully Rd (FL)	2	−35.368	138.638
E. ovata	Stipiturus CP (FL)	1	−35.371	138.551
Undetermined	Stipiturus CP (FL)	1	−35.371	138.551
E. paludicola	Stipiturus CP (FL)	1	−35.371	138.551
E. paludicola	Kelly Hill Caves CP (KI)	15	−35.997	136.873
E. paludicola	Short's property, original tree (KI)	1	−35.739	137.029
E. paludicola	Short's property, revegetation (KI)	9	−35.739	137.028
E. paludicola	Edwards Lagoon (KI)	4	−35.808	137.038
E. paludicola	O'Donnell property (KI)	5	−35.963	136.999
E. paludicola	Rocky River, natural (KI)	4	−35.946	136.753
E. paludicola	Rocky River revegetation (KI)	9	−35.947	136.741
E. paludicola	Nangkita (FL)	11	−35.345	138.711
E. paludicola	Burnfoot (FL)	8	−35.400	138.557
E. paludicola	Gold Diggings Swamp (FL)	2	−35.591	138.372
E. paludicola	Hindmarsh Valley (FL)	5	−35.402	138.518
E. paludicola	Range Rd (FL)	5	−35.567	138.469
E. paludicola	Parawa (FL)	2	−35.591	138.372
E. paludicola	Mosquito Hill Rd (FL)	19	−35.446	138.646
E. paludicola	Kokoda Rd (FL)	10	−35.407	138.688
E. paludicola	Cox Scrub CP (FL)	9	−35.331	138.747
E. paludicola	Proctor Road (FL)	7	−35.326	138.621

n = number of samples; KI, Kangaroo Island; FL, Fleurieu Peninsula.

We used a fusion PCR approach to generate an amplicon sequencing library for each individual [28]. Briefly, in a first step, we amplified in multiplex two panels of primers (24 primer-pairs each at 2 µM per primer in mix, supplementary Table S2), In a second step, we added a sample specific barcode by fusion PCR using the universal adapters added to the locus specific PCR primers as priming sites.

In more detail; for the first round of multiplex PCR we used the Multiplex PCR kit of Qiagen (Venlo, Netherlands) using the suggested manufacturers cycling conditions, with annealing temperature at 60 °C and amplifying 20 cycles to avoid overamplification. The reactions were done in 10 µL reactions adding 1 µL of primer mix (2 µM equimolar primer concentrations, supplementary Table S2) and 1.5 µL undiluted DNA template. Each primer consisted of the original sequence obtained from the *E. grandis* genome and an adapter sequence extension on the 5′ end to allow fusion PCR (*Eco*RI adaptor on forward and *Mse*I adapter on reverse, supplementary Table S4 in [25]). For the second round of PCR the same chemistry was used as in the first adding 1 µL *Eco*RI + CA and 1 µL *Mse*I + C fusion primers with internal barcodes (following supplementary Table S4 and Selective amplification in [25]) and 1µL of template (amplicons of first PCR). The reactions were done in 10 µL reaction with an annealing temperature of 60 °C and 15 cycles of amplification.

The resulting individual libraries were pooled, purified with AMPure XP (Agencourt, Beckman Coulter Inc., Brea, CA, USA) and then quantified using a 2200 TapeStation (Agilent, Santa Clara, CA, USA) with a high-sensitivity ScreenTape. The final library was diluted to 9.0 pMol and sequenced on an Ion PGM 318 v.1 chip. A total of 5 libraries were prepared to finalise the study.

2.4. Data Processing

The raw sequence data were imported into CLC Genomics Workbench 9 for demultiplexing, adapter and quality trimming (ambiguous trim = 2, quality limit = 0.05, minimum read length = 50). The resulting reads for each individual were then mapped to the reference sequences (indel and mismatch score = 2, similarity fraction = 1.0, length fraction = 0.5). With these mappings as input we used the fixed ploidy variant caller in CLC to identify SNPs (ploidy = 2, required variant probability = 0.95, minimum coverage = 10, minimum frequency = 0.2, filter homopolymer regions with minimum length = 3). Read mappings were then visually inspected and haplotypes were manually determined. For the final data we excluded loci that showed evidence of paralogy (more than 2 haplotypes per individual), amplified poorly (coverage <10 for >40% of individuals) or were invariant (minor allele frequencies <0.2). For all downstream data analyses, we treated each haplotype recovered as an allele for that locus.

2.5. Data Analyses

Summary statistics for species level genetic parameters including observed number of alleles (A) and allele frequencies, unbiased gene diversity (H_E), observed heterozygosity (H_O) and the fixation index (F_{IS}) were calculated using the software GenAlEx version 6.5 [29]. For these analyses, individuals were assigned to species groups based upon their morphological identification with the exception of those which, based upon the STRUCTURE results with K = 2, had an assignment probability to a cluster of <0.9.

Bayesian clustering analysis implemented in the program STRUCTURE 2.3.4 [30] was used to identify the most likely number of genetic clusters (K) among the samples and to assign individuals to clusters (input data in supplementary Table S3). Using this approach, individual genotypes can assign to a single cluster or can have mixed assignment when ancestry is shared in more than one parental group due to hybridisation (i.e., admixture). For these analyses, the admixture model with correlated allele frequencies with no priors of individual identification (i.e., morphological species assignment) were used. Ten independent runs at K values one to eight were run with MCMC simulations having 900,000 steps following a burn-in period of 100,000. The delta-K statistic [31] as implemented in STRUCTURE HARVESTER [32] was used to determine the most likely number of clusters (K) in the data. We assessed the average proportion of membership (Q_i) of samples to the inferred cluster and the individual membership proportion Q_i of each sample to the K clusters.

The methods implemented in NEWHYBRIDS 1.1 [33] were used to assess the evidence of hybrid ancestry for individual samples (input data in supplementary Table S3). NEWHYBRIDS uses MCMC simulations to provide a posterior probability of an individual assignment to predefined genealogical classes. We ran these analyses assuming 2 generations of hybridisation resulting in six genealogical classes: one class for each parental type as well as first generation (F_1), second generation (F_2) and 2 backcross ($F_1 \times$ parent 1; $F_1 \times$ parent 2) hybrid classes. All analyses were run without prior information regarding individual membership. Analyses were run over 200,000 steps following 50,000 burn-in using "Jeffery's like priors" for mixing proportions and allele frequencies. Three replicate analyses were performed to assess consistency across runs. We used a posterior probability of 0.9 as a threshold for assigning an individual to a specific genealogical class.

Following Nielsen et al. [34] we conducted simulations for our empirical data in order to assess the power of these markers to distinguish among genealogical classes and assess the range of q values expected for admixed individuals. HYBRIDLAB [34] was used to simulate pure parental genotypes using real data from reference individuals that could be unambiguously assigned (individual assignment probability >0.95) to a genetic cluster in our STRUCTURE analyses (above) with K = 2. We simulated 200 genotypes for each parental type that were then used to successively generate F_1, F_2 and two backcross genotype classes comprising 200 simulated multilocus genotypes per genealogical class. The simulated data where then analysed using STRUCTURE, with K set to 2, in order to assess the efficiency and accuracy of these analyses to identify admixed genotypes. Similarly, simulated genotypes were analysed in NEWHYBRIDS to assess the efficiency with which this approach could allocate simulated individuals to the correct genotype class. For both sets of analyses, settings are as described above for the real genotype data. Finally, a Principal Coordinates Analysis was performed in

GenAlEx to visualise the separation between species and individuals relative to simulated data using 50 simulated genotypes per genotype class.

3. Results

3.1. Molecular Data

A suite of 30 gene regions that were consistently amplifying with high coverage were used from a set of 48 primer pairs developed. These 30 gene regions generated adequate sequence data, containing multiple SNP loci, for downstream analyses. Of the eighteen loci deemed not adequate, 13 amplified poorly or not at all, 4 were invariant, and for one locus more than 2 alleles were apparent in some individuals suggesting paralogy. From the 30 useable marker gene regions, the proportion of missing data per locus averaged <4%, with a single locus having a maximum of c. 40% while the remaining loci generally fell below 10% of missing values. Missing values per individual were on average <10% (range 0–70%). A final dataset of 151 individuals was used for further analyses (supplementary Table S1).

The number of alleles per locus ranged from 2 to a total of 21 alleles (supplementary Table S1). The average number of alleles across the 30 loci was highest for *E. paludicola* (7.17), followed by *E. cosmophylla* (3.1) and *E. ovata* (2.3). This presumably reflects differences in sample size included in this study across taxa, as well as a broader geographic sampling of *E. paludicola*. However, *E. paludicola* was heterozygous at all loci (observed heterozygosity, 0.017–0.968, average = 0.664) while *E. cosmophylla* and *E. ovata* were monomorphic at 5 and 9 loci, respectively, and had average observed heterozygosity of 0.319 (range, 0–0.833) and 0.301 (range, 0–1), respectively. Both *E. cosmophylla* and *E. ovata* showed complete segregation at 8 loci (c. 30%), while there were no markers that were diagnostic for *E. paludicola*. *E. cosmophylla* and *E. ovata* were strongly diverged with an F_{ST} of 0.453, in line with intersectional comparisons within subg. *Symphyomyrtus* [35], while F_{ST} values for *E. paludicola* were 0.131 (*E. cosmophylla*) and 0.167 (*E. ovata*).

3.2. Admixture Analyses and Hybrid Class Assignment

The delta-K statistic for the admixture analyses performed using STRUCTURE was clearly optimal for K = 2 (ΔK = 855.9) with the next best grouping being four clusters (ΔK = 10.9) (Figure S2 in supplementary Table S4). When 2 groupings were assumed, individuals morphologically identified as *E. cosmophylla* or *E. ovata* were each unambiguously assigned to a single distinct cluster. Individuals identified as *E. paludicola* were partially assigned to both clusters with individual membership proportions $Q_i < 0.9$ (Figure 1). Significantly, when 3 clusters were assumed, the *E. paludicola* samples did not form a discrete group, as would be expected under the hypothesis that it is a distinct taxon (data not shown). Using simulated data with two parental and 4 hybrid categories, the average membership proportions for parental genotypes were >0.98 to a single K cluster while average assignment probabilities for each of the simulated hybrid classes was <0.91 (maximum individual assignment probability = 0.9 for *E. ovata* backcross) (supplementary Table S5). The Q_i values of the actual genotype data for *E. paludicola* fell within the range of values found for simulated hybrids and below the values of simulated parental genotypes (Figure 1) and it is reasonable to conclude that most *E. paludicola* individuals can be classified as admixed.

The analyses of the simulated genotype data using NEWHYBRIDS assigned virtually all samples to their expected genealogical class with posterior probabilities exceeding 0.95 (supplementary Table S5) and in the few cases with lower confidence there were no misclassified individuals (i.e., probabilities were apportioned among classes). This suggests that the empirical genotype data contain adequate signal to correctly assign individuals to parental and various hybrid categories. Analyses of the empirical data using NEWHYBRIDS (Figure 2) closely reflect the results of the admixture analyses above (Figure 1) and in particular, individuals identified as *E. cosmophylla* were unambiguously assigned to a pure parental class (posterior probability >0.99), as were all but one *E. ovata* sample. The admixed individuals (*E. paludicola* and one *E. ovata* individual) identified by STRUCTURE were assigned to a range of hybrid classes with most individuals being placed in a single category. Of the 126 *E. paludicola* individuals included in these analyses, 90 (*c.* 71%) were unambiguously assigned (posterior assignment probability >0.9, mean assignment probability >0.99) to an F_1 class and 15 individuals (*c.* 12%) were identified as F_2 hybrids. The three individuals (c. 2.5%) identified as backcrossed to *E. ovata* were associated with revegetation sites, as were c. 75% of the F_2 individuals. Eighteen individuals (*c.* 14%) could not be confidently placed within a single genotypic class (assignment probabilities <0.9) but were fractioned between pure *E. cosmophylla* (2 individuals), F_1, F_2 and backcross classes (Figure 2). Assignment probabilities for each genealogical class were summed across all individuals to give an expected number of *E. paludicola* plants in each category (supplementary Table S3 Figure S3). These analyses indicate the predominance of F_1 (~76%) and F_2 genotypes (~17%) and a relatively low incidence of backcrossing (~3.5%), particularly in the direction of *E. cosmophylla* (~1.9%). Figure 3 plots the first two principal coordinates of 2 analyses (data in supplementary Table S1), one with the real data (all eucalypts used in study) and the 50 simulated genotypes for each genealogical class. The PCoA analysis showed a clear separation of *E. ovata*, *E. cosomophylla* and *E. paludicola*, which is placed in the intermediate position between them. For the simulated genotype data, the F_1 and F_2 classes cluster with *E. paludicola*, while the backcross classes are intermediate between *E. paludicola* and each parental taxon. More than 30% of the variation is explained by these two coordinates, the majority being on the horizontal axis (Figure 3). This supports the most important signal in the data being the relative placement of *E. ovata*, *E. cosomophylla* and *E. paludicola* and aligns the latter with the F_1 hybrids as determined by simulations.

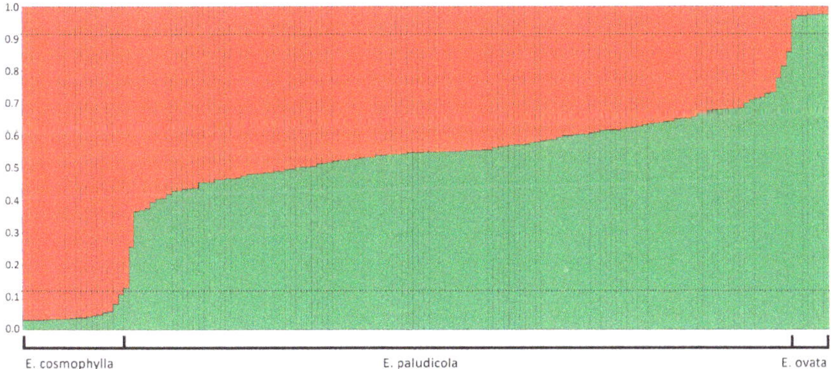

Figure 1. Proportion of ancestry for 151 *Eucalyptus* individuals with K = 2, inferred using STRUCTURE. Red and the green clusters correspond to *E. cosmophylla* and *E. ovata*, respectively, while *E. paludicola* shows admixture. The horizontal dashed lines indicate the range of Q values for simulated hybrid individuals (see text for details).

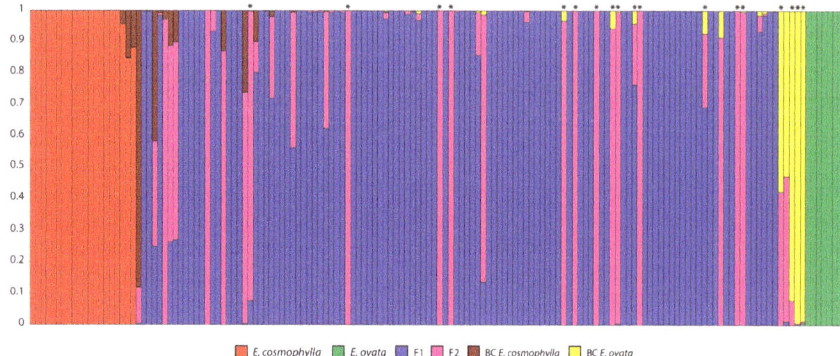

Figure 2. The proportion of individual assignment to 6 predefined genealogical classes for 151 *Eucalyptus* individuals, inferred using NEWHYBRIDS and 30 loci. Red and the green clusters correspond to "pure" *E. cosmophylla* and *E. ovata*, respectively, while *E. paludicola* is represented by a range of hybrid classes. The asterisks (*) correspond to *E. paludicola* individuals sampled from revegetation sites. BC in the figure stands for "back cross".

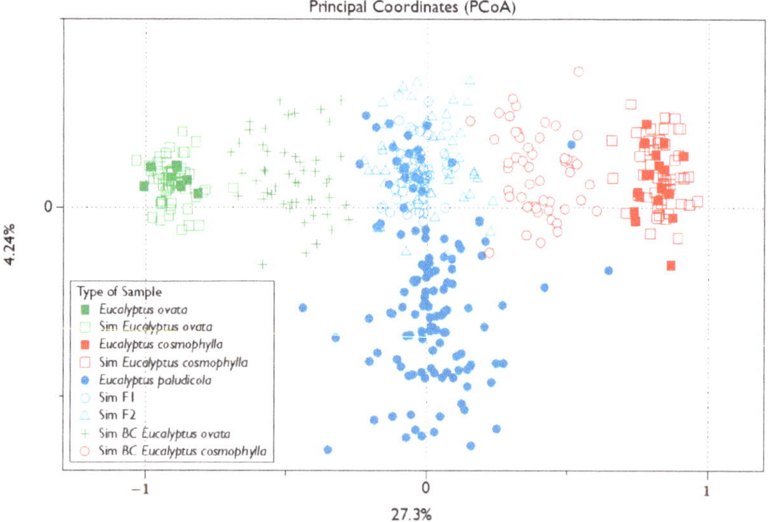

Figure 3. Principal Coordinates Analysis for real and simulated genetic data for *Eucalyptus paludicola*, *E. ovata*, *E cosmophylla*. First (F_1) and second (F_2) generation, back cross with *E. ovata* and back cross with *E. cosmophylla* were all simulated in HYBRIDLAB to depict the positioning of *E. paludicola*.

4. Discussion

The results of our study clearly indicate that *E. paludicola* is a hybrid taxon derived from *E. cosmophylla* and *E. ovata*, as has been previously suspected based upon morphological [18,23] and molecular evidence [21,22]. Despite the relatively high frequency of hybridisation among eucalypts, the formation of this hybrid would be unexpected given a relatively deep divergence between the progenitor species—*E. cosmophylla* and *E. ovata* are placed within different taxonomic sections of subg. *Symphyomyrtus*, which among eucalypts, strongly predicts the potential likelihood of successful hybrid formation [5,12,17]. Cases of natural intersectional hybridisation are relatively uncommon, and for example, Griffin et al. [5] reported intersectional hybrids within subg. *Symphyomyrtus* at a frequency of <5% amongst geographically proximal species. *E. paludicola* thus represents a rare case of natural intersectional hybridisation. A recent study based upon crossing experiments between *E. globulus* and congeners spanning a broad range of evolutionary distances suggests that complete reproductive isolation among eucalypts may take in excess of 20 million years to develop [17] (divergence time estimates based upon [36], but see Thornhill et al. [37], who infer significantly younger divergence times for eucalypt lineages based on a more densely sampled phylogeny). According to Crisp et al. [36] the divergence of section *Maidenaria* (Clade I of Steane et al. [38]; *E. ovata*) and the closest relations of section *Incognitae* (*E. cosmophylla*) that were included in their study (sect. *Exsertaria* and *Bisectae*, Clade II of Steane et al. [38]) occurred at between 5–15 million years before present, providing an approximate timescale for the divergence of *E. cosmophylla* and *E. ovata*.

These results indicate that the proposal for a possible instance of *Eucalyptus paludicola* as a hybrid speciation, so far unknown among eucalypts, is unsupported. To confirm this hypothesis, it would need to be supported by evidence that hybridisation has led to the development of reproductive isolation [2]. In addition to a probable hybrid origin, circumstantial evidence that *E. paludicola* is reproductively isolated from the parental taxa [22] has included limited variation in adult and seedling morphology throughout its geographical range, arguing against introgression and the observation that progeny planted in an arboretum show levels of morphological variation within the range of that taxon [18,22]. However, our findings suggest a contrasting interpretation of the status and origins of *E. paludicola*. In particular, the predominance of F_1 hybrids among *E. plaudicola* individuals included in this study (Figure 2) is consistent with a relatively uniform morphology that is intermediate to that of the parents but also with ongoing gene flow among the parental species, which occasionally results in hybrid formation. We have also found that hybridisation rarely proceeds beyond the F_1 stage, as indicated by the low frequency of F_2 and backcrossed individuals amongst our sample. This is despite the fact that many *E. paludicola* individuals are mature and produce viable seed, and herbarium collections observed at the State Herbarium of South Australia (AD) that are referred to *E. paludicola* date back to at least the 1920s, suggesting there has been opportunity for multiple generations of interspecific gene flow. Flowering asynchrony might limit the formation of later generation hybrids [39] although both parental species flower over a long period (autumn to spring i.e., April to November) that is coincident with *E. paludicola* (Spring i.e., August to November) [23] suggesting this is also unlikely. Interestingly, the majority of the F_2 and all of the unambiguously assigned backcrossed individuals were associated with the revegetation sites that were included in our study. The production of these individuals included seed germination and on-growing seedlings before planting in restoration sites enabling survival of F2 recruits. In contrast, the low frequency of post-F_1 hybrids among natural stands suggests low germination and/or survivorship under field conditions. Similar results have been found elsewhere among eucalypts, and for example, Shepherd and Lee [40] note an absence of mature post-F_1 interspecific *Corymbia* hybrids in the field, despite the apparent vigour and fecundity of F_1 individuals and their cultivated offspring.

Intrinsic outbreeding depression (OBD) provides a plausible explanation for the poor performance of hybrids reported in a number of controlled eucalypt crosses (e.g., [16,17,40–42]) and is manifested by factors including reduced seed viability, delayed germination and increased mortality throughout the life of a cohort. Compared to F_1s, recombination and segregation in later generation crosses may result in advanced generation hybrid breakdown due to disruption of coadapted gene complexes, or loss or duplication of chromosomal segments [4]. Minor disadvantages in performance of the hybrids between seed formation and maturity may be expressed by low survivorship and reproductive isolation can be maintained between parent species despite weak barriers to gene flow [43]. A recent study of *E. globulus* × *E. nitens* controlled crosses found that OBD may not be evident in advanced generations until age 2 C4 years and increased with age at least up to 14 years [16]. Late acting postzygotic isolation has also been implicated as a barrier to the formation of later generation hybrids in the field among eucalypts [40] and other long-lived tree species. e.g., [43]. Importantly, the likelihood and intensity of OBD is strongly predicted by evolutionary/taxonomic distance between eucalypt species [11,17]. OBD may provide a reproductive barrier between *E. ovata* and *E. cosmophylla*, which is suggested by:

(1) the globally low numbers of *E. paludicola* individuals, indicating that hybrid formation is infrequent; and (2) the high proportion of F_1 individuals within *E. paludicola*, consistent with selection against later generation hybrids.

Among eucalypts, the lack of evidence for hybrid speciation is perhaps surprising given their propensity for hybrid formation [22]. Eucalypts are remarkably constant in their base chromosome number [44–46] and this in part explains why so many natural hybrids occur, and why fertile synthetic interspecific hybrids are relatively easily obtained [45]. An implication of the above is that hybridisation leading to species formation would likely proceed without a change in chromosome number (i.e., homoploid hybrid speciation), a situation that has been considered vanishingly rare [2,47]. Hybrids can suffer reduced fertility and viability and may be intermediate in ecology and are therefore less fit than the parents in the parental habitat. Even in instances where the hybrid is fertile and locally adapted, weak barriers to gene flow between the hybrid and parental species present a substantial barrier to the development of reproductive isolation. The results of our study suggest that while reproductive isolation is incomplete, strong postmating isolation between *E. ovata* and *E. cosmophylla* is manifest by the production of a transient hybrid taxon, and there is no support for the recognition of a stable hybrid species.

Implications for Conservation

A hybrid origin for *E. paludicola*, which is currently listed under the EPBC Act as an endangered species, has implications for both its taxonomic and conservation status. Guidelines such as those outlined in the Biodiversity Conservation Act 2016 in the state of Western Australia [48] provide an objective basis to consider the status of *E. paludicola*. Under this scheme, hybrid entities must meet three criteria to warrant conservation listing: they are a distinct entity that is self-perpetuating and has arisen via natural processes. As a set of early generation, predominantly F_1 hybrids that have arisen multiple times via rare spontaneous events, we find little support for the ongoing conservation listing of *E. paludicola* or indeed recognition as a distinct species. A better conservation action would be to provide ongoing possibilities for the formation of the hybrid offspring of the two progenitor species, between *E. ovata* and *E. cosmophylla*.

The outcomes of hybridisation are diverse, ranging from the generation of new stable entities through to the loss of diversity through the erosion of species boundaries [1]. While hybridisation may have negative impacts on biodiversity, it has also been argued that hybrids arising from natural processes—and that are not a threat to the integrity of the parental species—should be afforded protection on the basis of preserving genetic diversity and natural evolutionary processes [49,50]. We detected a low level of backcrossing in the *E. paludicola* hybrid system suggesting that hybridisation is unlikely to present a threat to the demographic viability or genetic integrity of either parental species, both of which are relatively common in the region. On the other hand, it has been suggested that hybridisation may have potential adaptive benefits in the context of rapidly changing environments [8] and may have important ecosystem consequences (e.g., [51]). As a relatively rare example of natural intersectional hybridisation within *Eucalyptus*, *E. paludicola* is of scientific interest, for example, in unravelling the nature of reproductive isolation and speciation within the genus (e.g., [35]). However, it is difficult to justify active and ongoing management for *E. paludicola* beyond the maintenance of the parental species and their habitat. Several conservation listed eucalypt taxa have previously been found to be hybrids in light of molecular evidence (e.g., [6,9]). The results of our study further highlight the limitations of morphological evidence alone in delimiting conservation priorities [48] and in particular for identifying eucalypt hybrids [14] and their systems [17]. It points to the need of thorough molecular genetic analyses as part of any conservation assessment process where hybridisation is likely.

Supplementary Materials: The following are available online at http://www.mdpi.com/1424-2818/12/12/468/s1, Table S1: Samples and Genetic Diversity. Table S2: Primers developed for *Eucalyptus paludicola*. Table S3: Input data for the Structure and Newhybs analyses. Table S4: Structure and Newhybs final results. Table S5: NewHybrids simulated data.

Author Contributions: Conceptualization, M.W. and D.B. in consultation with the Department for Environment and Water *Eucalyptus paludicola* recovery team; methodology, K.-j.v.D., E.B., D.B., J.Q., H.C.; formal analysis, K.-j.v.D. and E.B.; investigation, E.B., M.W., K.-j.v.D., resources, M.W., D.B., J.Q.; data curation, K.-j.v.D. and M.W.; writing—original draft preparation, M.W., E.B. and K.-j.v.D.; writing—review and editing A.H.T., M.W., E.B. and K.-j.v.D.; visualization, K.-j.v.D.; supervision, M.W.; project administration, M.W., K.-j.v.D.; funding acquisition, M.W. All authors have read and agreed to the published version of the manuscript.

Funding: This research received was funded by the South Australian Department for Environment and Water (*Eucalyptus paludicola* recovery program) and The State Herbarium of South Australia.

Acknowledgments: Project support provided by Kylie Moritz (Natural Resources Adelaide and Mount Lofty and Natural Resources Murray Darlin Basin, Department for Environment and Water; current address Landscape South Australia—Murraylands and Riverland) for support with collection of samples, liaison with the *Eucalyptus paludicola* recovery team, advice on management actions. Technical support was provided by Ainsley Calladine (State Herbarium of South Australia), Martin O'Leary, samples were provided by Luke Price and Kym Ottewell. We thank the two anonymous reviewers whose comments helped improve the final manuscript.

Conflicts of Interest: The authors declare no conflict of interest.

References

1. Abbott, R.; Albach, D.; Ansell, S.; Arntzen, J.W.; Baird, S.J.E.; Bierne, N.; Boughman, J.; Brelsford, A.; Buerkle, C.A.; Buggs, R.; et al. Hybridization and speciation. *J. Evol. Biol.* **2013**, *26*, 229–246. [CrossRef] [PubMed]

2. Schumer, M.; Rosenthal, G.G.; Andolfatto, P. How common is homoploid hybrid speciation? *Evolution* **2014**, *68*, 1553–1560. [CrossRef] [PubMed]

3. Nicolle, D. Classification of the *Eucalypts* (*Angophora, Corymbia* and *Eucalyptus*) Version 4. 2019. Available online: https://dn.com.au/Classification-Of-The-Eucalypts.pdf (accessed on 10 December 2020).

4. Potts, B.M.; Dungey, H.S. Interspecific hybridization of *Eucalyptus*: Key issues for breeders and geneticists. *New For.* **2004**, *27*, 115–138. [CrossRef]

5. Griffin, A.R.; Burgess, I.P.; Wolf, L. Patterns of natural and manipulated hybridisation in the genus *Eucalyptus* L'Herit—A review. *Aust. J. Bot.* **1988**, *36*, 41–66. [CrossRef]

6. Rossetto, M.; Lucarotti, F.; Hopper, S.D.; Dixon, K.W. DNA fingerprinting of *Eucalyptus graniticola*: A critically endangered relict species or a rare hybrid? *Heredity* **1997**, *79*, 310–318. [CrossRef]

7. Stokoe, R.L.; Shepherd, M.; Lee, D.J.; Nikles, D.G.; Henry, R.J. Natural Inter-subgeneric Hybridization Between *Eucalyptus acmenoides* Schauer and *Eucalyptus cloeziana* F. Muell (Myrtaceae) in Southeast Queensland. *Ann. Bot.* **2001**, *88*, 563–570. [CrossRef]

8. Field, D.L.; Ayre, D.J.; Whelan, R.J.; Young, A.G. Molecular and morphological evidence of natural interspecific hybridization between the uncommon *Eucalyptus aggregata* and the widespread *E. rubida* and *E. viminalis*. *Conserv. Genet.* **2009**, *10*, 881–896. [CrossRef]

9. Walker, E.; Byrne, M.; Macdonald, B.; Nicolle, D. Clonality and hybrid origin of the rare *Eucalyptus bennettiae* (Myrtaceae) in Western Australia. *Aust. J. Bot.* **2009**, *57*, 180–188. [CrossRef]

10. Delaporte, K.L.; Conran, J.G.; Sedgley, M. Interspecific Hybridization within *Eucalyptus* (Myrtaceae): Subgenus *Symphyomyrtus*, Sections *Bisectae* and *Adnataria*. *Int. J. Plant. Sci.* **2001**, *162*, 1317–1326. [CrossRef]

11. Potts, B.; Barbour, R.; Hingston, A.B.; Vaillancourt, R. Genetic pollution of native eucalypt gene pools—Identifying the risks. *Aust. J. Bot.* **2003**, *51*, 1–25. [CrossRef]

12. Dickinson, G.R.; Wallace, H.M.; Lee, D.J. Reciprocal and advanced generation hybrids between *Corymbia citriodora* and *C. torelliana*: Forestry breeding and the risk of gene flow. *Ann. For. Sci.* **2013**, *70*, 1–10. [CrossRef]

13. Schuster, T.M.; Setaro, S.D.; Tibbits, J.F.G.; Batty, E.L.; Fowler, R.M.; McLay, T.G.B.; Wilcox, S.; Ades, P.K.; Bayly, M.J. Chloroplast variation is incongruent with classification of the Australian bloodwood eucalypts (genus *Corymbia*, family *Myrtaceae*). *PLoS ONE* **2018**, *13*, e0195034. [CrossRef] [PubMed]

14. Ellis, M.F.; Sedgley, M.; Gardner, J.A. Interspecific Pollen-Pistil Interaction in *Eucalyptus* L'Hér. (Myrtaceae): The Effect of Taxonomic Distance. *Ann. Bot.* **1991**, *68*, 185–194. [CrossRef]

15. Gore, P.L.; Potts, B.M.; Volker, P.W.; Megalos, J. Unilateral cross-incompatibility in *Eucalyptus*: The case of hybridisation between *E. globulus* and *E. nitens*. *Aust. J. Bot.* **1990**, *38*, 383–394. [CrossRef]

16. Costa e Silva, J.; Potts, B.M.; Tilyard, P. Epistasis causes outbreeding depression in eucalypt hybrids. *Tree Genet. Genomes* **2012**, *8*, 249–265. [CrossRef]

17. Larcombe, M.J.; Holland, B.; Steane, D.A.; Jones, R.C.; Nicolle, D.; Vaillancourt, R.E.; Potts, B.M. Patterns of reproductive isolation in *Eucalyptus*—A phylogenetic perspective. *Mol. Biol. Evol.* **2015**, *32*, 1833–1846. [CrossRef]

18. Nicolle, D. A new series, *Incognitae*, of *Eucalyptus* L'Hér., including a new species endemic to Fleurieu Peninsula and Kangaroo Island, South Australia. *J. Adel. Bot. Gard.* **1995**, *16*, 73–78.

19. T.S.S.C. Commonwealth Listing Advice on *Eucalyptus paludicola*. 2006. Available online: http://www.environment.gov.au/biodiversity/threatened/species/pubs/eucalyptus-paludicola-advice.pdf (accessed on 10 December 2020).

20. Quarmby, J. *Action Plan for Mount Compass Swamp Gum (Eucalyptus paludicola) 2011–2016*; Government of South Australia, Department of Environment and Natural Resources: Adelaide, Australia, 2016; p. 17.

21. Ottewell, K.; Bickerton, D.; Quarmby, J.; Lowe, A.J. *Genetic Status of the Endangered Mount Compass Swamp Gum (Eucalyptus paludicola)*; University of Adelaide, Department of Environmental and Natural Resources: Adelaide, Australia, 2011.

22. Jones, R.C.; Nicolle, D.; Steane, D.A.; Vaillancourt, R.E.; Potts, B.M. High density, genome-wide markers and intra-specific replication yield an unprecedented phylogenetic reconstruction of a globally significant, speciose lineage of *Eucalyptus*. *Mol. Phylogenet. Evol.* **2016**, *105*, 63–85. [CrossRef]

23. Nicolle, D. Myrtaceae (partly) (version 1). In *Flora South Australia*, 5th ed.; State Herbarium of South Australia: Adelaide, Australia, 2014; p. 102.

24. Myburg, A.A.; Grattapaglia, D.; Tuskan, G.A.; Hellsten, U.; Hayes, R.D.; Grimwood, J.; Jenkins, J.; Lindquist, E.; Tice, H.; Bauer, D.; et al. The genome of *Eucalyptus grandis*. *Nature* **2014**, *510*, 356–362. [CrossRef]

25. Cross, H.; Biffin, E.; van Dijk, K.-J.; Lowe, A.; Waycott, M. Effective application of next-generation sequencing (NGS) approaches in systematics and population genetics: Case studies in *Eucalyptus* and *Acacia*. *Aust. Syst. Bot.* **2016**, *29*, 235–246. [CrossRef]

26. Kearse, M.; Moir, R.; Wilson, A.; Stones-Havas, S.; Cheung, M.; Sturrock, S.; Buxton, S.; Cooper, A.; Markowitz, S.; Duran, C.; et al. Geneious Basic: An integrated and extendable desktop software platform for the organization and analysis of sequence data. *Bioinformatics* **2012**, *28*, 1647–1649. [CrossRef] [PubMed]

27. Untergasser, A.; Cutcutache, I.; Koressaar, T.; Ye, J.; Faircloth, B.C.; Remm, M.; Rozen, S.G. Primer3—New capabilities and interfaces. *Nucleic Acids Res.* **2012**, *40*, e115. [CrossRef] [PubMed]

28. O'Neill, E.M.; Schwartz, R.; Bullock, C.T.; Williams, J.S.; Shaffer, H.B.; Aguilar-Miguel, X.; Parra-Olea, G.; Weisrock, D.W. Parallel tagged amplicon sequencing reveals major lineages and phylogenetic structure in the North American tiger salamander (*Ambystoma tigrinum*) species complex. *Mol. Ecol.* **2013**, *22*, 111–129. [CrossRef] [PubMed]

29. Peakall, R.; Smouse, P.E. GENALEX 6: Genetic analysis in EXCEL. Population genetic software for teaching and research. *Mol. Ecol. Notes* **2006**, *6*, 288–295. [CrossRef]

30. Pritchard, J.K.; Stephens, M.; Donnelly, P. Inference of population structure using multilocus genotype data. *Genetics* **2000**, *155*, 945–959. [PubMed]

31. Evanno, G.; Regnaut, S.; Goudet, J. Detecting the number of clusters of individuals using the software STRUCTURE: A simulation study. *Mol. Ecol.* **2005**, *14*, 2611–2620. [CrossRef]

32. Earl, D.; von Holdt, B. STRUCTURE HARVESTER: A website and program for visualizing STRUCTURE output and implementing the Evanno method. *Conserv. Genet. Resour.* **2012**, *4*, 359–361. [CrossRef]

33. Anderson, E.C.; Thompson, E.A. A Model-Based Method for Identifying Species Hybrids Using Multilocus Genetic Data. *Genetics* **2002**, *160*, 1217–1229.

34. Nielsen, E.E.; Bach, L.A.; Kotlicki, P. HYBRIDLAB (version 1.0): A program for generating simulated hybrids from population samples. *Mol. Ecol. Notes* **2006**, *6*, 971–973. [CrossRef]

35. Hudson, C.J.; Freeman, J.S.; Myburg, A.A.; Potts, B.M.; Vaillancourt, R.E. Genomic patterns of species diversity and divergence in *Eucalyptus*. *New Phytol.* **2015**, *206*, 1378–1390. [CrossRef]

36. Crisp, M.D.; Burrows, G.E.; Cook, L.G.; Thornhill, A.H.; Bowman, D.M.J.S. Flammable biomes dominated by eucalypts originated at the Cretaceous–Palaeogene boundary. *Nat. Commun.* **2011**, *2*, 193. [CrossRef] [PubMed]

37. Thornhill, A.H.; Crisp, M.D.; Külheim, C.; Lam, K.E.; Nelson, L.A.; Yeates, D.K.; Miller, J.T. A dated molecular perspective of eucalypt taxonomy, evolution and diversification. *Aust. Syst. Bot.* **2019**, *32*, 29–48, 20. [CrossRef]

38. Steane, D.A.; Nicolle, D.; Sansaloni, C.P.; Petroli, C.D.; Carling, J.; Kilian, A.; Myburg, A.A.; Grattapaglia, D.; Vaillancourt, R.E. Population genetic analysis and phylogeny reconstruction in *Eucalyptus* (Myrtaceae) using high-throughput, genome-wide genotyping. *Mol. Phylogenetics Evol.* **2011**, *59*, 206–224. [CrossRef] [PubMed]

39. Barbour, R.C.; Potts, B.M.; Vaillancourt, R.E.; Tibbits, W.N. Gene flow between introduced and native *Eucalyptus* species: Flowering asynchrony as a barrier to F1 hybridisation between exotic *E. nitens* and native Tasmanian *Symphyomyrtus* species. *For. Ecol. Manag.* **2006**, *226*, 9–21. [CrossRef]

40. Shepherd, M.; Lee, D.J. Gene flow from *Corymbia* hybrids in northern New South Wales. *For. Ecol. Manag.* **2016**, *362*, 205–217. [CrossRef]

41. Lopez, G.A.; Potts, B.M.; Tilyard, P.A. F1 hybrid inviability in *Eucalyptus*: The case of *E. ovata* × *E. globulus*. *Heredity* **2000**, *85 Pt 3*, 242–250. [CrossRef]

42. Volker, P.W.; Potts, B.M.; Borralho, N.M.G. Genetic parameters of intra- and inter-specific hybrids of *Eucalyptus globulus* and *E. nitens*. *Tree Genet. Genomes* **2008**, *4*, 445–460. [CrossRef]

43. Lindtke, D.; Gompert, Z.; Lexer, C.; Buerkle, C.A. Unexpected ancestry of *Populus* seedlings from a hybrid zone implies a large role for postzygotic selection in the maintenance of species. *Mol. Ecol.* **2014**, *23*, 4316–4330. [CrossRef]

44. Grattapaglia, D.; Bradshaw, H.D., Jr. Nuclear DNA content of commercially important *Eucalyptus* species and hybrids. *Can. J. For. Res.* **1994**, *24*, 1074–1078. [CrossRef]

45. Oudjehih, B.; Abdellah, B. Chromosome numbers of the 59 species of *Eucalyptus* L'Herit. (Myrtaceae). *Caryologia* **2006**, *59*, 207–212. [CrossRef]

46. Grattapaglia, D.; Vaillancourt, R.E.; Shepherd, M.; Thumma, B.R.; Foley, W.; Külheim, C.; Potts, B.M.; Myburg, A.A. Progress in Myrtaceae genetics and genomics: *Eucalyptus* as the pivotal genus. *Tree Genet. Genomes* **2012**, *8*, 463–508. [CrossRef]

47. Yakimowski, S.B.; Rieseberg, L.H. The role of homoploid hybridization in evolution: A century of studies synthesizing genetics and ecology. *Am. J. Bot.* **2014**, *101*, 1247–1258. [CrossRef] [PubMed]

48. Moritz, C. Conservation units and translocations: Strategies for conserving evolutionary processes. *Hereditas* **1999**, *130*, 217–228. [CrossRef]

49. Allendorf, F.W.; Leary, R.F.; Spruell, P.; Wenburg, J.K. The problems with hybrids: Setting conservation guidelines. *Trends Ecol. Evol.* **2001**, *16*, 613–622. [CrossRef]

50. Jackiw, R.N.; Mandil, G.; Hager, H.A. A framework to guide the conservation of species hybrids based on ethical and ecological considerations. *Conserv. Biol.* **2015**, *29*, 1040–1051. [CrossRef] [PubMed]

51. Dungey, H.S.; Potts, B.M.; Whitham, T.G.; Li, H.-F. Plant genetics affects arthropod community richness and composition: Evidence from a synthetic eucalypt hybrid population. *Evolution* **2000**, *54*, 1938–1946. [CrossRef]

Publisher's Note: MDPI stays neutral with regard to jurisdictional claims in published maps and institutional affiliations.

Article

Genetic Patterns and Climate Modelling Reveal Challenges for Conserving *Sclerolaena napiformis* (Amaranthaceae s.l.) an Endemic Chenopod of Southeast Australia

Michael D. Amor *, Neville G. Walsh and Elizabeth A. James

Royal Botanic Gardens Victoria, South Yarra 3141, Australia; neville.walsh@rbg.vic.gov.au (N.G.W.); elizabeth.james@rbg.vic.gov.au (E.A.J.)
* Correspondence: michael.amor@rbg.vic.gov.au

Received: 14 October 2020; Accepted: 2 November 2020; Published: 4 November 2020

Abstract: *Sclerolaena napiformis* is a perennial chenopod endemic to southeast Australia. Human-mediated habitat loss and fragmentation over the past century has caused a rapid decline in abundance and exacerbated reduced connectivity between remnant populations across three disjunct regions. To assess conservation requirements, we measured the genetic structure of 27 populations using double digest RADseq). We combined our genetic data with habitat models under projected climate scenarios to identify changes in future habitat suitability. There was evidence of regional differentiation that may pre-date (but also may be compounded) by recent habitat fragmentation. We also found significant correlation between genetic and geographic distance when comparing sites across regions. Overall, *S. napiformis* showed low genetic diversity and a relatively high proportion of inbreeding/selfing. Climate modelling, based on current occupancy, predicts a reduction in suitable habitat for *S. napiformis* under the most conservative climate change scenario. We suggest that the best conservation approach is to maximise genetic variation across the entire species range to allow dynamic evolutionary processes to proceed. We recommend a conservation strategy that encourages mixing of germplasm within regions and permits mixed provenancing across regions to maximise genetic novelty. This will facilitate shifts in genetic composition driven by individual plant fitness in response to the novel environmental conditions this species will experience over the next 50 years.

Keywords: Camphorosmoideae; conservation genetics; disjunct distribution; population fragmentation; population structure; single nucleotide polymorphism (SNP) markers

---

1. Introduction

The International Union for the Conservation of Nature (IUCN) recently listed >1250 Australian species as being vulnerable to extinction, which places the risk to Australia's biodiversity as amongst the highest globally [1]. Currently, genetic factors are not used for IUCN listings, but their inclusion can improve the assessment of extinction risk and subsequent conservation of some species [2]. Developers of biodiversity conservation policy acknowledge the need to retain dynamic evolutionary processes [3] and thus the security of those processes remains a major goal for the management of genetic resources [4,5] including the conservation of population genetic diversity and individual fitness [6,7]. Reproductive strategies, including the breeding system and dispersal capacity of pollen and seed, influence the distribution of genetic diversity, which can have a strong effect on plant population fitness [8]. For many plant species, predictions of increased aridity and shifting seasonality compound the effects of recent human-mediated changes to their habitat [9]. These factors combine to escalate the extinction risk of threatened species by reducing habitat connectivity, thereby disrupting patterns of gene-flow and potentially reducing effective population size and increasing levels of inbreeding [10].

Plant species respond to local changes in environmental conditions at the population level [11], ensuring ongoing homogeneity in genetic diversity across the landscape is increasingly important [12]. Ideally, spatial and temporal monitoring of intraspecific variation would inform species management where human impacts are likely to have negative impacts on phenotypic and genetic variation [13]. Collectively, evolutionary processes, high levels of heterozygosity [14], and allelic diversity [15] are understood to enable adaptation and resilience to environmental change; thus, conserving natural patterns of genetic variation and interaction should be prioritised to ensure species' longevity [5,16]—particularly for small plant populations subject to environmental stress [14]. If dispersal is insufficient for plants to track changing climate [17], human-assisted migration beyond the natural dispersal envelope may be justified [18]. Determining the amount and distribution of variation within a species and the minimum population size required to retain evolutionary potential are critical components in conservation planning [19,20], but the quantification of these factors remains elusive for many species.

The alluvial plains of the southern Riverina area of New South Wales and Victoria and the Victorian Wimmera riverine area of southeastern Australia [21] have been acknowledged for their productivity, both in cropping and livestock raising, since being established for European agriculture. The progressive introduction of a large-scale irrigation network from the late 1800s saw extensive conversion of the native shrublands and open woodlands to broad-scale agriculture, resulting in extreme reduction and fragmentation of the original vegetation. Subsequently, the overall community in this region was listed as Critically Endangered [22]—the direct result of weed incursion, salination from rising water tables, uncontrolled grazing, and human-induced climate change with associated drying and altered fire regimes [23]. Documentation of the original vegetation of this area (e.g., [24,25]) indicates that chenopods, particularly *Atriplex*, *Maireana*, and *Sclerolaena*, formed a significant component of the low shrub and ground layer.

The reduction in chenopod species in riverine habitats has led to the recommendation that ex situ germplasm and genetic restoration involving plant translocations should be undertaken to ensure their conservation in riverine habitats. Importantly, these actions are tempered by the phylogeographic history and distribution of intraspecific variation. Determining whether distinct genetic lineages exist within taxa [26] is necessary to avoid the unintended introduction of poorly adapted plants [27,28]. New, cryptic chenopod taxa are still being described from the region [29,30], which highlights a risk of accidental inclusion of non-target species in restoration actions. When planning these strategies, it is important to consider that artificially restricting the gene pool may have the unintended consequence of limiting population adaptability [31,32].

A chenopod of conservation concern in southeastern Australia, *Sclerolaena napiformis* Wilson (Turnip Copperburr) [33], is one of 66 species of *Sclerolaena*, an endemic genus restricted to mainland Australia. *Sclerolaena* is a member of the subfamily Camphorosmoideae within Amaranthaceae s.l., which includes the previously recognised families Amaranthaceae s.s. and Chenopodiaceae [34]. Australian Camphorosmeae comprises a monophyletic lineage of ≈150 perennial C3 species in contrast to the largely C4-dominant Amaranthaceae s.s [35]. The distribution of Australian plant species prior to colonisation is generally poorly documented, and due to large-scale habitat conversion, many extant species are believed to inhabit only a minute proportion of their former range [36]. It is likely that *S. napiformis* was widely distributed within its current extent of occurrence but through fragmentation is now confined to three disjunct areas. Extant populations are largely restricted to roadsides, travelling stock routes, and small reserves within an otherwise highly modified landscape [23]. The historical patterns of gene-flow within the metapopulation are now greatly disrupted and the species is considered nationally endangered [22].

There are two major concerns regarding the future of *S. napiformis*. One is the likely effect of population decline and loss of connectivity on the genetic health of the species. The second is the reduction in the availability of suitable habitat caused by changing climate and habitat conversion. Site degradation, a contributor to habitat loss, occurs primarily from weed invasion by species either better adapted to changing conditions and/or that can outcompete *S. napiformis* (e.g., overtopping by dense foliage of the non-native grasses *Phalaris aquaticum* and *Lophopyrum ponticum*). Some conservation work has been undertaken for the species, including translocation of plants into secure landholdings and augmentation of populations on road reserves. However, this work has been undertaken without knowledge of the distribution of genetic diversity across the geographic range of *S. napiformis*. A national recovery plan for the species [23] advocates in situ protection and the establishment of an ex situ seed collection, but provides no recommendation for assessing genetic attributes that may pinpoint important populations or guide further actions such as translocation, assisted migration, and genetic rescue.

In this study, we used a reduced representation genomic sequencing method (double digest restriction-site associated DNA sequencing-ddRADseq; [37]) to examine genetic patterns by comparing single nucleotide polymorphism (SNP) loci within and among populations across the geographic range of *S. napiformis*. We aimed to quantify the genetic structure and the partitioning of variation within *S. napiformis* at the population and regional scale, as the presence of genetic structure could influence the sourcing of germplasm for restoration. We also looked for the presence of allelic variants restricted to populations or regions and identified chloroplast haplotypes on the basis of ddRAD SNP loci. Finally, we modelled the future suitability of *S. napiformis* habitat using climate and soil variables. Using this combined approach, we recommend procedures for the collection and use of germplasm to refine the conservation strategy for *S. napiformis*.

2. Materials and Methods

2.1. The Study Species

Sclerolaena napiformis is restricted to three disjunct regions in the semi-arid Murray Darling Depression and Riverina regions of Victoria (Vic) and New South Wales (NSW) in southeast Australia. We designated the following regions: "northeast" (NSW only), where only two populations were found; "central", with populations north and south of the Murray River; and "southwest" (Victoria only), containing the greatest number and density of populations (Figure 1). The species grows as a procumbent multi-stemmed, perennial sub-shrub up to 40 cm high, mainly in native grassland and remnant Buloke (*Allocasuarina luehmannii*) woodland habitats. A thick rootstock develops within one year of germination and plants can re-shoot from the stem base following dieback in response to environmental conditions [23]. The floral morphology of *S. napiformis* is typical of wind pollination, and the fruiting perianth, a burr, has spines, suggesting dispersal by animals (zoochory) (Figure 2). The breeding system has not been studied but seed is produced if plants are bagged to exclude pollen from a different individual; thus, we assume *S. napiformis* to be self-compatible (James, unpublished data), as apomixis has not been recorded in the genus. At a few sites, the species commonly associated with at least low levels of salinity are present (e.g., *Atriplex semibaccata*, *Salsola tragus* subsp. *tragus*, *Spergularia brevifolia*), but overall, the associated vegetation carries little signal of salinity. All collection sites comprised primarily native species where weeds may have been present but did not dominate the site, suggesting that *S. napiformis* is not tolerant of significant modification of the original vegetation.

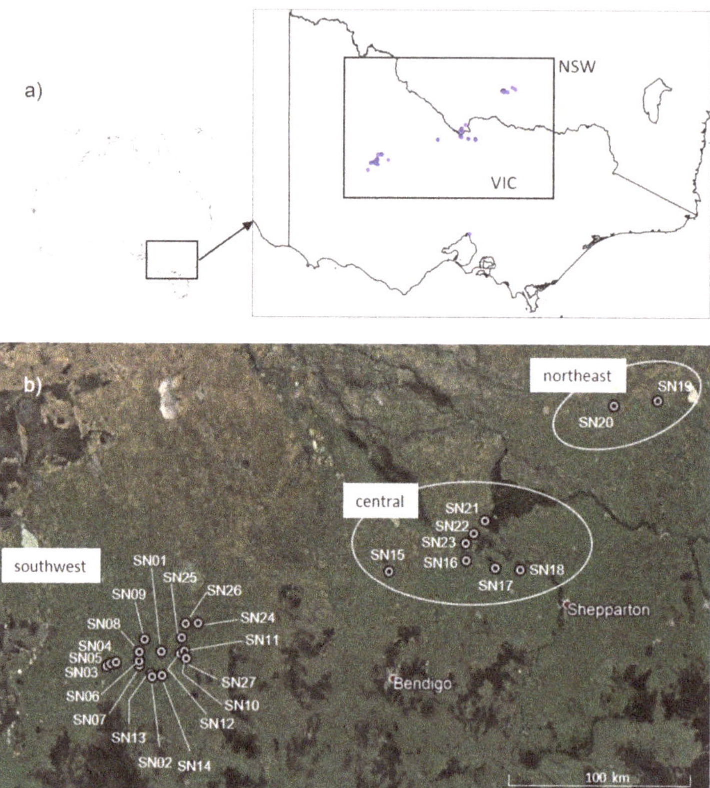

Figure 1. Distribution of *Sclerolaena napiformis* and location of sampled sites. (**a**) Geographic distribution based on herbarium records, indicated by blue dots, downloaded from Australasia's Virtual Herbarium (AVH 2019). (**b**) Location of the three disjunct regions with sampled sites labelled.

Figure 2. Habit, flowers, and fruit of *Sclerolaena napiformis*. (**a**) A thick rootstock develops within one year of germination and plants can re-shoot from base of stem following dieback. Developing fruit circled. The insignificant flowers, lack of petals, and exposure of stigmas and anthers to air currents are features typical of wind pollination. (**b**) Young flower with two stigmas (♀) and five unopened anthers (♂, two underdeveloped) visible. (**c**) An older flower with three stigmas (♀) and three dehisced anthers visible (♂). (**d**) Female reproductive organs excised from fruit showing stigma lobes (♀) and a single ovary (O). The timing of stigma receptivity is not known. (**e**,**f**) Developing fruit, with remnant of stigmas still visible (**f**). (**g**) Mature dry fruit collected from stems. Circular scar at base of one fruit shows the position of fruit attachment to the stem.

2.2. Sampling

For population analysis (ddRADseq and chloroplast haplotying), we visited 44 sites from February to April in 2018 (late summer to mid-autumn) following identification of sites using location information taken from herbarium specimens in the Australasian Virtual Herbarium (avh.ala.org.au) and observation records in the Atlas of Living Australia (ala.org.au). No plants were found at 15 sites and populations at two sites had insufficient leaf tissue to sample. The 27 sites sampled across the species' geographic range varied in area from 0.06 to 1.25 ha and contained discrete (sub)populations varying in size from approximately 40 to 400 individuals. Plants were sometimes cryptic, and thus these estimates, which were based on observed plants, were generally conservative. Sampled plants were geolocated and fresh leaf material was collected from 9–22 mature plants per population and desiccated in silica gel (Figure 1, Table 1).

Table 1. Site codes, region, sample numbers, estimated area, and number of plants per site (see Figure 1 for location of sites within regions).

Site	Region	Number of Samples Collected	Estimated Area (ha)	Estimated Pop Size
SN01	Southwest	9	0.6	50
SN02	Southwest	25	1.0	380
SN03	Southwest	20	0.3	65
SN04	Southwest	21	0.06	40
SN05	Southwest	20	0.15	340
SN06	Southwest	20	1.25	200
SN07	Southwest	20	1.8	80
SN08	Southwest	20	0.56	150
SN09	Southwest	20	1.0	150
SN10	Southwest	20	0.8	170
SN11	Southwest	20	0.45	210
SN12	Southwest	20	3.0	120
SN13	Southwest	20	1.125	150
SN14	Southwest	20	0.8	130
SN15	Central	20	0.7	400
SN16	Central	20	1.2	250
SN17	Central	23	0.11	120
SN18	Central	19	1.14	350
SN19	Northeast	12	0.44	70
SN20	Northeast	23	0.24	500
SN21	Central	20	1.05	500
SN22	Central	20	1.5	100
SN23	Central	20	5.0	80
SN24	Southwest	6	0.05	40
SN25	Southwest	23	1.25	120
SN26	Southwest	21	1.8	120
SN27	Southwest	19	0.9	180

In addition, two samples, grown from seed collected from Trevaskis Rd (SN18, central region) and South Corree Rd (SN20, northeast region), were used for whole genome Illumina shotgun sequencing to provide a genome scaffold for ddRADseq SNP loci. Chloroplast reference genomes [38] were assembled from the same sequence data and used for haplotyping.

2.3. Chromosome Counts

To assist in the analysis of ddRADseq data, we determined the ploidy of *S. napiformis* by chromosome counts. Actively growing root tips were collected from ex situ plants grown from seed collected at populations SN12, SN18, SN20 (at or very near the site of the type collection), and SN24. Root tips were prepared using a modification of the method of Murray and Young [39] by pre-treatment in a saturated 1,4-dichlorobenzene solution at 20 °C for 24 h, then fixation in 3:1 95% ethanol/glacial acetic acid at 4 °C for 24 h. Root tips were rinsed several times in water and either stored in 70% ethanol at −20 °C or prepared immediately for counts by hydrolysing root tips in 10% HCl for 8 min at 60 °C, macerated in a drop of FLP orcein stain [40], then squashed and viewed under oil immersion. Chromosome counts were made from cells across several root tips per plant.

2.4. DNA Isolation and Sequencing

Leaf material was dried to maximise DNA preservation during extensive field surveys. Dried leaf material (≈20 mg) was ground to a fine powder in a QIAGEN TissueLyser II (two minutes) and genomic DNA was isolated using the CTAB protocol for ISOLATE II Plant DNA Kit (Bioline)—except that final elution was in a total volume of 40 μL elution buffer. DNA quality was confirmed via 1% agarose gel electrophoresis in 1xTBE buffer for 45 min at 100 V and stained with SybrSafe (Invitrogen, Thermo Fisher Scientific Australia). DNA isolations were quantified using a Qubit v3.0 fluorometer (Invitrogen Life Technologies) and stored at −20 °C.

2.4.1. Whole Genome Sequencing

To improve identification of loci from the ddRADseq reads, we obtained a de novo reference scaffold using Illumina shotgun sequences. Genomic DNA (CTAB protocol for ISOLATE II Plant DNA Kit; Bioline Australia) was isolated separately from fresh leaf material sampled and vouchered from 2 cultivated plants grown using seed sourced from Jerilderie, New South Wales (MEL2470835A) and near Wyuna, Victoria (MEL2446779A). A genome library was prepared for each sample using a TruSeq Nano DNA Library Prep kit (Illumina, San Diego, USA). A minimum of 2.5 μg genomic DNA template of each sample was supplied to Australian Genome Research Facility (AGRF; Parkville, Aust) at 50 ng/μL and sequenced (150 bp PE on an Illumina NovaSeq 6000).

2.4.2. ddRADseq

A modified version of the Peterson et al. ddRADseq protocol [37] was used to prepare DNA libraries—details are available at michaelamor.com/protocols. The final DNA library contained 480 samples, which included 13 technical replicate pairs. Individuals were randomly allocated to a plate per well and therefore received a random barcode and index. In summary, 100 ng of genomic DNA was digested for 18 h with EcoRI-HF and AseI restriction enzymes (New England Biolabs) and barcoded adapters were ligated to digested DNA fragments. Non-ligated adapters were removed before libraries were size-selected by magnetic bead purification using Jetseq Clean (Bioline)/PEG 8000 buffer solution at 0.5× then 0.9× of the DNA solution volume.

PCR-based indexing of the individual libraries was conducted using real-time PCR (rtPCR) in a Bio-Rad CFX96 thermocycler, with amplification stopped after 10 cycles. To assess whether amplification was adequate for sequencing whilst minimising PCR bias, we visualised individual fluorescence curves to ensure that they had not plateaued. Amplified libraries were pooled in equal concentrations based on relative fluorescence unit outputs from rtPCR and were concentrated/purified using Jetseq Clean beads/PEG 8000 buffer solution (1.8× DNA solution volume). The pooled library was size-selected at 300–500 bp in a Pippin Prep (Sage Science) using a 2% agarose (100–600 bp) cassette and quantified via qPCR using a Jetseq Library Quantification Hi-ROX kit (Bioline) on a Bio-Rad CFX96 thermocycler. Library QC and sequencing were performed at the AGRF.

2.5. Quality Filtering and Bioinformatics

2.5.1. Nuclear DNA Assembly

Genome scaffolds for each sample were assembled as follows. Quality filtering of reads was performed using Trimmomatic 0.39 [41] with the following settings: LEADING:3 TRAILING:3 SLIDINGWINDOW:4:20 MINLEN:36. Paired reads were then normalised to ≈100× coverage using BBnorm (settings: mindepth = 5 target = 100). Normalised reads were assembled into scaffolds using

SPAdes v3.13.0 [42]. Benchmarking Universal Single-Copy Orthologs (BUSCO v3.1.0; [43]) statistics were applied to assess the completeness of the assembly. To estimate the recovery of gene-coding regions in the resulting assembly, we assessed scaffolds with BUSCO using the embryophyta_odb10 database (settings: -m genome -sp arabidopsis).

2.5.2. Population Samples (ddRADseq)

Raw paired-end reads were trimmed if the quality dropped below a score of phred20, based on a sliding-window of four bases using Trimmomatic v0.38 [41]. Reads below our minimum length requirement of 100 bases were discarded. Finally, reads were trimmed if Illumina adapters were present and filtered for microbial and fungal contaminants using Kraken v2.0.6 [44]. Reads were demultiplexed into individual sample read sets using the "process_radtags" feature of STACKS v1.4.6 [45].

Assembly of reads into loci was performed on (i) reads 1 and 2, and (ii) only read 1 using ipyrad v0.7.29 [46]. Reads were assembled using a stepwise approach by mapping reads to our most reliable genome scaffold (determined by size (gb) and BUSCO score) and performing de novo assembly on the remaining reads. Inclusion of read 1 only resulted in more loci in all cases, and therefore assemblies based only on read 1 were used downstream. Further sequence quality filtering was performed to convert base calls with a score of <30 into Ns, whilst excluding reads with ≥15 Ns. Our complete assemblies required minimum depth of 10x reads per locus. The resulting assemblies were used to generate variant call format (VCF) files for population genomic analyses and alignment (phylip) files for chloroplast (cp) DNA-based haplotyping.

The occurrence of outlier loci, interpreted as candidates for loci under selection, was assessed for all assemblies using Bayescan v2.1 [47] and a Principal Component Analysis (PCA) approach implemented via the "pcadapt" package [48] in R 3.5.3 [49]. Only sites identified by both methods were considered true outliers. BayeScan (10,000 pilot runs and 200,000 generations, with 50,000 initial generations discarded as burn-in) identified zero outliers compared to 104 outliers detected by "pcadapt". As no outliers were common across both approaches, we considered those identified by "pcadapt" to be false positives, and the following analyses were conducted using all loci.

2.5.3. Population Samples (cp Haplotyping)

Reads from the ddRADseq libraries were mapped to a consensus sequence of two *S. napiformis* individuals (GenBank Accessions MT027236 and MT027237 [38]) using ipyrad [46]. The clustering similarity threshold was set to a minimum of 80% and the presence of ≥50 individuals per locus. An alignment including the two reference genomes was visualised in Geneious Prime v2020.1.2 (https://www.geneious.com). A summary table of variable sites was created manually, and a haplotype network was constructed in R 3.5.3 using the "pegas" package [50].

2.6. Analysis of Genetic Diversity and Structure

Genetic structure was investigated among and within regions using a step-wise approach via the "fastStructure" algorithm [51]. VCF files generated by ipyrad were converted to ped and map files using VCFtools v0.1.15 [52], then further converted to bed, bim, and bam files using PLINK v1.90b4 [53]. To identify genetic clusters, we performed 8 to 10 replicate runs for each K-value (number of sites minus 1) based on our complete assembly using the logistic model with 10 cross-validation steps for each run. FastStructure outputs were summarised with Clustering Markov Packager Across K (CLUMPAK) "main pipeline – admixture" and "best K" algorithms via the online server (http://clumpak.tau.ac.il/). We attempted 150 fastStructure replicate runs for each K-value for our regional analysis, however, a high failure rate resulted in us retaining >10 replicates for K = 3 but only two completed replicates for both K = 2 and K = 4. Due to low (and uneven) replicate numbers, we were unable to summarise using CLUMPAK. Therefore, regional analyses were summarised using fastStructures with "chooseK.py" and "distruct.py" scripts to enable presentation of the single "best" graphical representation and its associated likelihood values. To investigate potential sub-structure between the two northeast sites, we performed fastStructure analyses using an assembly generated from individuals from the northeast and central sites. We compared the Bayesian-based genetic clusters obtained from fastStructure with those identified using K-means clustering and Bayesian information criterion (BIC) as implemented for discriminant analysis of principal components (DAPC [54]). To support the above findings, we investigated correlation between genetic and geographic distance using Mantel's R statistic via the "mantel" function in the "vegan" R package [55]. Geographic (Euclidean) and genetic distance (dissimilarity) matrices were calculated with the "dist" and "bitwise.dist" functions in R, respectively, using the package "poppr" [56].

For population genomic level analyses within and among regions, we imported VCF files into R and converted them to "genind" and "hierfstat" objects. Estimates of genetic differentiation were calculated via pairwise G_{ST} and Jost's D among our identified regions using the "mmod" R package. The degree of genetic differentiation among regions was estimated via 1000 permutations of analysis of molecular variance (AMOVA) using the "poppr" R package. Euclidean genetic distance was input as a covariate using the "pegas" algorithm. Global and regional estimates of genetic diversity were calculated based on the observations resulting from our population structure analyses. We estimated observed heterozygosity (Ho) and gene diversity (He) using the "basic.stats" function in the "hierfstat" R package. Global F_{IS} was estimated via the "boot.ppfis" function in the "hierfstat" R package [57] to determine the proportion of species-wide and region-wide genetic variation reflected in individuals. Global F_{ST} and G_{ST} was estimated via the "wc" and "Gst_Hedrick" functions in "hierfstat" and "mmod" [58], respectively.

2.7. Distribution Modelling

We investigated the potential for habitat/environment suitability to shift under a conservative climate change scenario using 20 variables (Table S1). A total of 8 soil variables (3 seconds resolution) were downloaded from the CSIRO soil and landscape database (http://www.clw.csiro.au/aclep/soilandlandscapegrid/index.html). A total of 12 bioclimatic variables (30 seconds resolution) were downloaded from the WorldClim database (www.worldclim.org/bioclim) as "current conditions" (interpolations of observed data, representative of 1960–1990) and "future conditions" (IPCC5 climate projections for 50 years assuming the most conservative "representative concentration pathway" (RCP = 2.6) for greenhouse gases). Collection coordinates were used as species occurrence data, which is the most extensive collection to date and was informed by >30 years of field surveys.

The extent of the models was set to 10 degrees surrounding occurrences (site locations). Each model was fitted using the maximum entropy (Maxent) algorithm via the *dismo* package in R [59]. Model fit was evaluated by simulating 1000 random pseudoabsences within the model extent and tested how well our collection data compared to "random guessing" using a receiver operating characteristic (ROC) curve (AUC = 0.994; Figure S1a). We considered this to be a realistic approach to obtaining absences, as true absence data are difficult to obtain for *S. napiformis* because plants can die down to ground level under stressful conditions [23]. Variable contributions were plotted for (i) current climate, (ii) soil, and (iii) current climate and soil combined (Figure S1b). We used the raster package in R [60] to predict the suitable habitat for *S. napiformis* in 2050 using our "future climate" variables and including soil data with the assumption that soil and landscape will remain constant.

3. Results

3.1. Chromosome Counts

All counts were consistently 2n = 18 chromosomes (Figure 3). A diploid number of 2n = 18 was reported for *Sclerolaena birchii* (recorded as *Bassia birchii*, [61]), the only other species of *Sclerolaena* for which chromosome numbers are available. Therefore, we considered *S. napiformis* diploid, and ploidy was specified as such where required for analysis.

Figure 3. Root tip squash showing 2n = 18 chromosomes for an individual of *Sclerolaena napiformis* grown from seed collected at Cope Cope, Victoria (site SN24), viewed under oil immersion (×1000). Scale bar = 10 μm.

3.2. Data assembly Statistics

Our nuclear genome assembly (built using short reads) was ≈6.5 million nucleotides in length and consisted of 733,636 contigs (average length = 888.27, largest length = 154,606). Overall, 1145 complete genes were identified (83% of the 1375 total recognised genes), 137 genes (10%) were partially identified (fragmented sequence), and 93 (6.7%) were missing.

For ddRAD data, the final analysed dataset consisted of 452 individuals. Summary statistics are provided in Table 2. After filtering loci and informative sites for each assembly, the number of unlinked SNPs ranged from 2837 to 3367 with a mean depth of ≈15 reads per individual at every retained locus. For all assemblies, approximately 31% of reads were mapped to our assembled genome scaffold.

Table 2. Summary statistics for *Sclerolaena napiformis* assemblies generated using ipyrad. Statistics reported for "unlinked SNPs" and "mean locus depth" were obtained after filtering variant call format (VCF) files with VCFtools. SNP: single nucleotide polymorphism.

	All Regions	Southwest	Central	Northeast
No. individuals in assembly	452	296	126	30
Total loci	3827	4043	5169	5011
Informative sites	5518	4971	3659	4082
Percent reads mapped to reference	31.2	31.3	31.3	30.0
Mean error rate of base calls (± SD)	0.003 (0.001)	0.003 (0.001)	0.003 (0.001)	0.003 (0.001)
Unlinked SNPs	3093	3045	2837	3367
Mean locus depth (± SD)	15.31 (8.83)	15.21 (17.24)	15.65 (7.20)	15.37 (9.93)
Mean individuals per locus (± SD)	163.38 (43.45)	108.11 (28.63)	46.64 (12.39)	12.17 (3.59)

3.3. Regional Genetic Diversity and Structure

Measures of regional genetic diversity are summarised in Table 3. Genetic structure among regions was supported by AMOVA (see Tables S2–S5), where 8.7% of the total variation was observed among southwest, central, and northeast regions (p = <0.001). AMOVA also provided support for differentiation among sites within the southwest and northeast regions (p = <0.02), but not among sites within the central region (p = 0.8). Correlation between genetic and geographic distance (Mantel's R) was significant among regions (p = 0.001), and among sites within the central region (p = 0.04). There were no noteworthy correlations among sites within the southwest or northeast region. Overall, *S. napiformis* was characterised by relatively low levels of heterozygosity (Ho = 0.017), with a high proportion of the overall genetic diversity captured within a given individual (F_{IS} = 0.617), pointing to notable levels of inbreeding/selfing within the species. This trend was consistent when looking within regions where heterozygosity levels (Ho) were relatively low but diversity within individuals (F_{IS}) was high. All sites exhibited F_{IS} values consistent with substantial levels of selfing, except SN24, where the F_{IS} value was negative (Table S6), suggesting clonal reproduction was greatest at this site. There was no correlation between F_{IS} and population size, population area, or the number of samples included in the estimation of F_{IS}.

Table 3. Measures of genetic diversity (observed heterozygosity (Ho) and gene diversity (He)) and fixation indices, calculated to determine the proportion of genetic variation contained in sites, relative to the total observed variation (F_{ST} and G_{ST}) and the proportion of the overall variation contained in an individual (F_{IS}). Analyses were performed on the entire dataset ($n = 452$). N = number of sites, n = number of samples. Statistics based on assemblies of each region were also calculated. Mantel's R statistic is also reported with significant correlation between genetic and geographic distances marked by an asterisk.

	N	n	Ho	He	F_{ST}	G_{ST}	F_{IS}	Mantel's R (*p*-Value)
All regions	27	452	0.017	0.056	0.156	0.209	0.617	0.088 (0.001) *
Southwest	18	296	0.027	0.058	0.205	0.304	0.517	−0.009 (0.646)
Central	7	126	0.040	0.112	0.225	0.288	0.652	0.103 (0.038) *
Northeast	2	30	0.049	0.258	0.200	0.369	0.830	0.106 (0.153)

Analyses among the three disjunct regions showed that sites within the southwest, central, and northeast all had greater gene diversity than heterozygosity and were characterised by a high degree of inbreeding/selfing (F_{IS} = 0.517, 0.652, and 0.830, respectively). Population differentiation among sites was similar within the southwest and central regions (G_{ST} = 0.304 and 0.288, respectively) but higher for the small northeast region (G_{ST} = 0.369). Comparing pairwise genetic difference between regions, we found the central and northeast regions to be the most distinct (pairwise G_{ST} = 0.228, Jost's D = 0.017; Table 4), despite being geographically closer than either was to the southwest.

Table 4. Pairwise genetic differentiation (lower left: G_{ST}, upper right: Jost's D) among three regions of *Sclerolaena napiformis* from Victoria and New South Wales, Australia. We report the greatest difference occurring between central and northeast regions.

Region	Northeast	Central	Southwest
Northeast	-	0.017	0.014
Central	0.228	-	0.010
Southwest	0.180	0.189	-

3.4. Population Genetic Structure

3.4.1. Overview of All Regions

Population structure was examined for the species across the geographic range. Marginal likelihoods resulting from fastStructure runs of the regional assembly of *S. napiformis* favoured four distinct genetic clusters with clear differences shown among geographic regions (Figure 4a). For K = 4, individuals assigned to blue and purple clusters were widely dispersed geographically but most common in the southwest region, whereas individuals from green and yellow clusters were mostly found in the central and northwest regions, respectively. DAPC identified the optimal number of clusters as three (Figure S2), and also showed differences among groups of individuals on the basis of their geographic region, but samples from the northeast were less tightly clustered than the majority of samples from the central and southwest regions (Figure 4b).

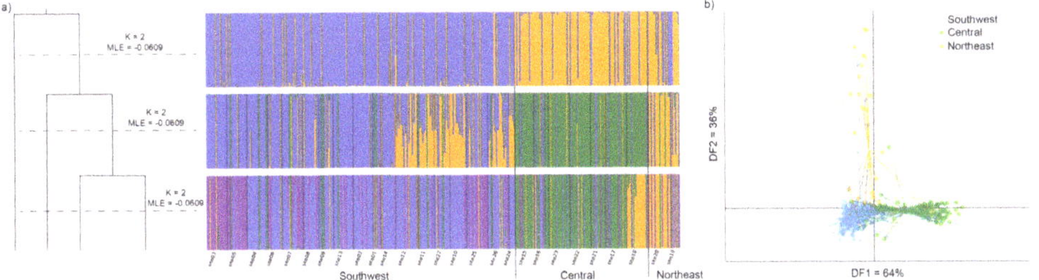

Figure 4. Group assignment and clustering outputs produced via (**a**) fastStructure showing marginal likelihoods and individual assignment to genetic clusters for K = 2–4, and (**b**) discriminant analysis of principal components (DAPC). K = 3 highlights distinction among southwest, central, and northeast regions and is consistent with fastStructure where some southwest populations show admixture with the yellow cluster dominating the northeast. Analyses were based on 3093 unlinked double digest RADseq SNPs obtained from an ipyrad assembly of 452 *Sclerolaena napiformis* individuals. Individuals are coloured according to region. No a priori locality information was input into either analysis.

3.4.2. Central and Northeast Regions

For the combined central and northeast regions, the optimal number of clusters identified from both fastStructure and DAPC was K = 4, although K = 7 was also a possibility (Figure S2). The distribution of clusters was not always well correlated with geographic location. Visualising the fastStructure output (Figure 5), at K = 4, the two northeast sites (19 and 20) were delimited from other sites but could not be distinguished from each other (Figure 5a). At K = 7, three groups were evident, with site 18 (central) and sites 19 and 20 (northeast) distinct from all other sites, which were dominated by a single genetic cluster (blue—Figure 5b). The remaining 1–4 clusters (K4 vs K7) were made up by very few individuals dispersed seemingly randomly across localities. DAPC was less successful at delimiting sites within the combined central and northeast region (Figure S3).

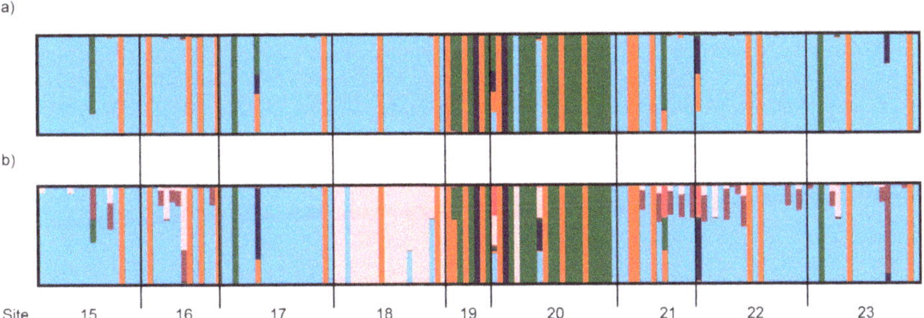

Figure 5. Clumpak output summarising 10 replicate fastStructure runs of (**a**) K = 4 and (**b**) K = 7 for *Sclerolaena napiformis* individuals from central (sites 15, 16, 17, 18, 21, 22, and 23) and northeast (sites 19 and 20) regions combined.

3.4.3. Southwest Region

The optimal number of clusters within the southwest region as identified by fastStructure was K = 4 (Figure 6), with populations showing variable levels of admixture. There was a loose geographic association with the more westerly sites being mainly cluster 1 (blue) and more easterly sites (10, 11, 12, 24, 26, and 27) being mainly cluster 2 (purple). Sites 5 and 25 had high membership probability to cluster 3 (green) that was not explained by habitat or proximity. Cluster 4 (orange) showed no geographic association. Although not definitive, DAPC suggested six genetic clusters as the lowest likely value of K and no substantial geographic structure with most individuals assigned to a single widespread genetic cluster (Figure S4).

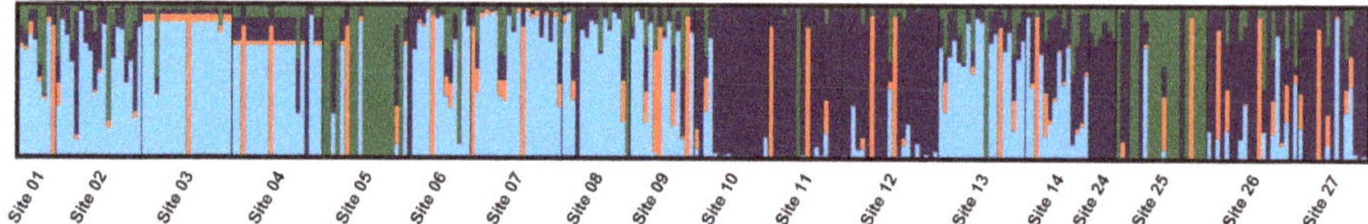

Figure 6. Clumpak output summarising eight replicate fastStructure runs for K = 4, which was optimal across a range of K2–K18. Genetic structure is evident between *Sclerolaena napiformis* sites to the east (sites 1–9 and 13–14) and west (sites 10–12 and 24–27) of Marnoo, Victoria.

3.5. Chloroplast Genomes and Haplotyping

We obtained 17 loci comprising 7 informative sites by mapping the ddRADseq reads to two chloroplast reference genomes (GenBank accessions MT027236 and MT027237). None of the informative sites mapped to the most variable chloroplast regions identified in the alignment of the reference genomes. Two informative sites corresponded to partial sequences of the matK gene, a variable site when comparing GenBank sequences for other species of *Sclerolaena*. The remaining informative sites did not correspond to any chloroplast sequences available for *Sclerolaena*. Although eight haplotypes were recognised, only six were differentiated if variable positions containing "N" were ignored, and while 59% of populations had more than one haplotype, it was not possible to identify relationships to sites or regions (Table 5). The star-shaped haplotype network (Figure S5a) is consistent with a single maternal lineage dominated by one haplotype, H1, and a few minor variants derived by mutation [62,63]. Haplotype H1, possibly ancestral, dominated overall and was recovered from 351/390 individuals spread across the 27 sites. The second most common haplotype, H2 (21/390 individuals), was found at 10/27 sites (37%) and in most individuals (9/14) at site 25, the only site where H1 was not the most common haplotype (Figure S5b).

Table 5. Haplotypes recovered from mapping ddRADseq reads to two reference *Sclerolaena napiformis* cpDNA genomes, SN20 and SnTR1 (151,530 bases long, 814 variable sites in total between the references). Six haplotypes were differentiated using unambiguous variable sites recovered in 390 individuals. The position of variable sites is reported relevant to the reference alignment (available via Mendeley Data; http://dx.doi.org/10.17632/664sh75jgz.1) and whether the variation occurs within a gene/coding region.

Haplotype		Variables Sites						
		125,670 -	125,801 -	134,268 rpoC1 Gene	134,367 rpoC1 Gene	134,376 rpoC1 Gene	115,143 matK CDS	115,263 matK CDS
	n							
H1	351	A	C	G	C	G	T	A
H2	21	T	C	G	C	G	T	A
H3	5	A	T	A	C	G	T	C
H4	1	A	T	G	C	G	T	C
[1] H4A	8	A	T	G	C	N	T	C
[1] H5A *	1	A	C	G	C	A	N	N
[1] H5B *	2	A	C	N	C	N	C	C
H6	1	A	C	G	T	G	T	A
Reference:								
SN20	1	A	C	G	C	G	T	A
TR01	1	A	T	G	C	G	T	C

[1] Haplotype 4A collapses to haplotype 4 due to unresolved nucleotide at position 134,376; * haplotypes 5A and 5B collapse into a single haplotype (H5) due to unresolved nucleotides at positions 134,376, 115,143, and 115,263. When compared only at ddRADseq loci, Sn20 and SnTR1 were equivalent to H1 and H4, respectively.

3.6. Distribution Modelling

Investigating the variables most associated with the occurrence of *S. napiformis* revealed that most variation in the model was accounted for by the presence of clay in the soil (≈50%) and the annual mean temperature (≈15%). Carbon and other climate variables associated with precipitation and temperature all contributed >5% towards the total observed variation (Figure S1b). Overall, a decrease in available suitable environment was predicted for the next 50 years using both models: climate variables only and a combination of climate and soil variables. Combining soil types with climate variables decreased predicted present and future suitable environment substantially when compared to the climate-only model. Additionally, the climate-only model highlighted a more southerly area that increased in suitability at approximately −38 degrees latitude. Including soil type almost entirely excluded this area, as the presence of clay soils played a major role in predicting present day occurrences (Figure 7).

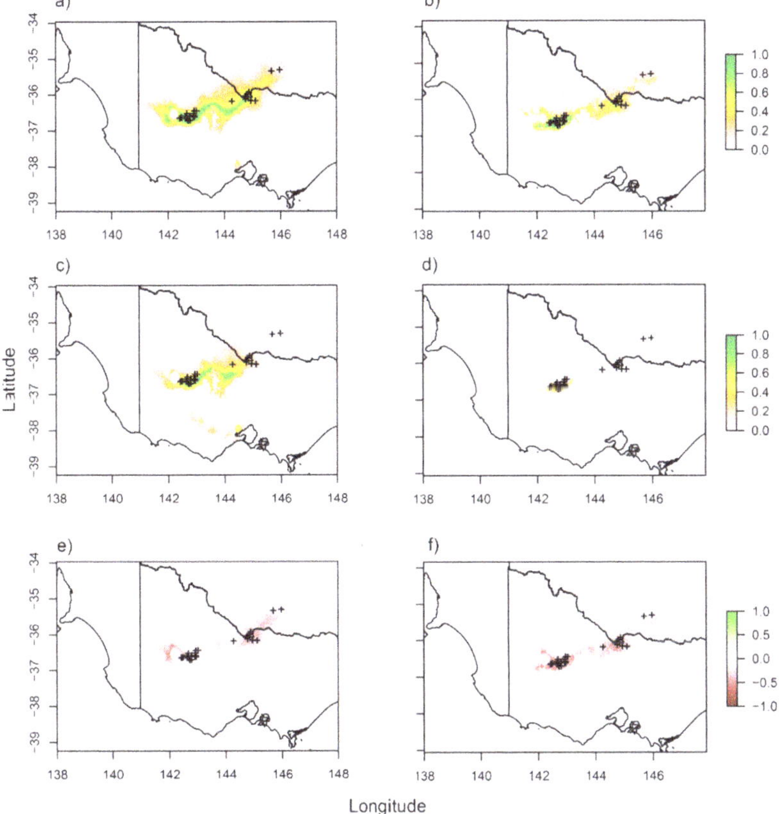

Figure 7. Predicted suitable habitat for *Sclerolaena napiformis* based on collections performed during this study. Modelling was based on (**a**) current climate, (**b**) current climate and soil composition, (**c**) future climate, (**d**) future climate and current soil composition. We also show a shift in climate suitability (**e**) excluding and (**f**) including current soil variables (increasing suitability shown by green and decreasing shown by red). Modelling was performed using 12 bioclimatic and 8 soil composition variables (Table S1). Future climate predictions are based on a conservative climate scenario (RCP 2.6) projected 50 years.

4. Discussion

We provide evidence for genetic structure among three disjunct regions of *S. napiformis* in southeast Australia. We did not find genetic differentiation among sites within central and northeast regions, but clear genetic structure was evident in the southwest region (east and west of Marnoo, Victoria). Our findings of regional differentiation of *S. napiformis* are consistent with limited long-distance dispersal (likely impacted by range disjunction), but also point to some historical dispersal beyond extant populations. Lower than expected heterozygosity, as well as high F_{IS} values, also indicates substantial inbreeding within populations of *S. napiformis*, a self-compatible, wind-pollinated species. Effective gene-flow is unlikely to be maintained across the distribution of this species, particularly given that human-mediated landscape changes have created barriers to gene-flow and access to suitable habitat. Even if pollen-mediated gene-flow is maintained among existing sites, the probability of new sites being colonised is low without human intervention. Importantly, we did not find geographically correlated cryptic lineages of *S. napiformis*, in contrast to findings for other disjunct species (e.g., [26]). We suggest that conservation management of *S. napiformis* utilises the genetic diversity found in all populations from the northeast, central, and southwest (divided into east and west sites) to maximise the adaptability of the species in increasingly novel environments.

We also show that the current distribution of *S. napiformis* is restricted by soil type and tightly linked to temperature and precipitation. As such, climate projections suggest that over half the existing habitat will become unsuitable for *S. napiformis* over the next 50 years. Whilst we predict a decline in habitat suitability across the entire species distribution, our projections suggest that populations in the southwest region are the most likely to persist under future climate projections. Overall, our findings highlight the complexities of managing the competing demands of human utilisation and conservation of native species in the face of climate change, but we outline some viable options below.

4.1. Fragmentation, Genetic Structure, and Gene-Flow

We recorded a single dominant cpDNA haplotype across the distribution of *S. napiformis*, which is consistent with historical broad-scale seed dispersal, although low coverage obtained via our non-cpDNA targeted approach may limit our ability to distinguish finer-scale patterns [64]. Our findings based on nuclear DNA show clear genetic differentiation among regions and highlight finer-scale structure in the southwest region. Therefore, dispersal in *S. napiformis* appears most common at relatively small scales (i.e., within 20–80 km), despite seeds having morphological adaptations that promote long-distance dispersal by birds and mammals. Relatively localised dispersal is considered common among plants [65,66], although genetic studies confirming this assumption are lacking [67]. Here, we provide valuable genetic evidence that seed dispersal does not maintain genetic homogeneity over the entire (≈370 km) linear range of *S. napiformis*.

Human activity (land clearing and fragmentation) has reduced suitable/available grassland habitat in this area by as much as 99% [23], which has clearly contributed to the isolation of extant populations and sites. However, we identified two genetic clusters in the southwest region, which were separated by ≈11 kms and displayed no obvious landscape barrier. As many of our sites are isolated by greater distances, we believe proximity alone cannot account for reduced connectivity among *S. napiformis* populations, despite extreme alteration to the landscape. Similarly, a sympatric species, *Pimelea spinescens*, was considered to have genetic structure that pre-dated fragmentation resulting from agriculture post-colonisation [68], as also observed in other plant systems [69,70]. We therefore suggest that the genetic structure observed in *S. napiformis* (northeast, central, and southwest—east and west division) was present prior to colonisation, although recent land conversion will likely have downstream consequences on genetic patterns.

Certainly, habitat loss and fragmentation displace biodiversity and biomass, and this is particularly problematic for grassland systems [71]. There is no doubt that the extreme alteration to the landscape has displaced organisms that historically assisted dispersal in *S. napiformis*. For example, the endangered plains-wanderer *Pedionomus torquatus* consumes the seed and leaves of *Sclerolaena* spp. [72]. Despite a small home range of 7–21 ha, this bird may have historically assisted in seed dispersal among sites when both *P. torquatus* and *S. napiformis* were more common. Ants also assist in the short-distance dispersal of a closely related (and sometimes co-occurring) species, *Sclerolaena diacantha* [73,74], although this has not been shown explicitly for *S. napiformis*. Furthermore, vehicles and grazing stock may contribute to the dispersal of this common "roadside" species, although we believe these roadside populations are more likely to reflect the remaining suitable habitat for this species.

4.2. Genetic Diversity and Inbreeding

We observed high levels of inbreeding and homozygosity in discrete *S. napiformis* regions (and 26/27 sites), which suggests a high degree of self-pollination. Indeed, a closely related species, *S. diacantha*, has an estimated selfing rate of 70% with strong selection for outcrossed progeny considered unlikely [73]. *Sclerolaena napiformis* may display similar selfing rates to *S. diacantha*, given that these species largely co-occur and thus experience similar environmental pressures. Mixed mating systems, whereby plants undergo both selfing and outcrossing, can minimise the accumulation of deleterious alleles [75], even following inbreeding depression [76] or under novel or changing environments [77]. This mechanism may be used by *S. napiformis* as a short-term counterbalance to the negative fitness consequences associated with selfing. It may also enable recruitment into new sites from a single or few related individual/s [13,69]. For example, the inbreeding species *Geum urbanum* showed high levels of homozygosity and selfing without any negative consequences for fitness [78]. We note, however, that negative genetic consequences associated with a reduced gene pool may still develop over a longer timeframe [13] and, considering the extreme habitat degradation in this system and the potential establishment of populations from few individuals, we believe there is a strong requirement for further genetic monitoring of *S. napiformis*.

4.3. Considerations for Persistence of S. napiformis

Our modelling suggests that climate change will substantially decrease the suitability of remaining habitat for *S. napiformis*, even under the most conservative projections. As the climate warms, organisms are required to adapt and/or shift their distribution to track suitable climate envelopes [79]. However, most plants are unlikely to migrate fast enough to keep pace with the current rate of climate change [17]. Furthermore, the reliance of *S. napiformis* on clay soil compounds this, as suitable soil is rare outside of its current distribution. We therefore believe it is unlikely for *S. napiformis* to undergo a range-shift naturally, leaving it exposed to increasingly unsuitable environmental conditions. Furthermore, climate change is predicted to favour an increase in weediness, particularly within an agricultural setting [80]. We expect this to further reduce habitat quality for *S. napiformis* as competition with invasive species increases. Therefore, we consider the persistence of *S. napiformis* into the future unlikely without human intervention, even if an effort is made to ensure remnant patches are protected. Active restoration and rehabilitation of converted land would thus go a long way towards ensuring the persistence of *S. napiformis* and the grasslands of southeast Australia.

The seeds of *S. napiformis* tolerate a wide range of temperatures, similar to other congeneric taxa (e.g., *Sclerolaena bicornis* [81] and *S. birchii* [61]). They are produced in high volume; may remain viable while dormant for several years [82,83]; and germinate rapidly once the physical barrier of the persistent, woody perianth is removed [81,84]. Furthermore, germination of *S. napiformis* seed is linked to sufficient rainfall and removal of the perianth, rather than being tied to a specific time of year [61,85]. This strategy of "germinating rapidly when most advantageous" may be favourable when rainfall is erratic, and might explain why *S. napiformis* alters its phenology to avoid water stress at sensitive stages of its life cycle [85]. *Sclerolaena napiformis* also appears to flower and reproduce year-round, and again, this is partly determined by the availability of water (personal observation). This finding, and the above-mentioned germination triggers, support the notion that water availability, rather than temperature, may have a greater influence on *S. napiformis* establishment and successful persistence upon encountering a new site. For example, drought sensitivity is known to limit plant distribution and recruitment, particularly if it impacts early-life stages [86]. Recruitment into new areas is likely to be more restricted than germination because of the additional sensitivity during seedling establishment; however, the volume of seeds contained in the natural soil seedbank allows for mass germination when conditions are suitable [84,87]. This greatly increases the likelihood of establishing a viable population in a new area once a small number of recruits reach maturity.

Finally, we observed that the northeast region contained the lowest number and poorest quality of mature plants in terms of plant size and seed production. However, we observed good survival and healthy growth of northeast seedlings cultivated under optimal ex situ conditions. Throughout the distribution of *S. napiformis*, reduced available habitat, altered fire regimes, hotter/drier conditions with decreasing precipitation, and increased biomass of weeds threaten populations with local extinctions. This is particularly likely in the northeast region, and therefore retention of the genetic diversity in this area is particularly important as plants here may harbour adaptations to the predicted future conditions for central and southwest regions. We believe all populations of *S. napiformis* will require assisted restoration in the future, and therefore consider seed-banking from each region (particularly the northeast) a conservation priority.

5. Conclusions and Implications for Conservation

For *S. napiformis* to persist, conservation efforts must balance the rate of local extinction (measured as population loss) with the colonisation of new sites. Successful conservation will rely partly on management strategies that facilitate connectivity among *S. napiformis* sites and consider the projected effect of climate change when identifying candidate localities. High fecundity of mature plants and long-term storage capacity of germplasm will benefit conservation, and even though the development of an ex situ seedbank is underway, continued collections from diverse populations over multiple time periods is recommended.

Small populations are more vulnerable if their genetics and evolutionary biology are not integrated into the conservation strategy [88]. Although most remaining populations of *S. napiformis* have <200 individuals, they also have an unusually high proportion of reproductive plants [2], which lessens the risk of diversity loss in translocated or artificial populations [89]. Both inbreeding and outbreeding depression are potential risk factors when combining germplasm or augmenting gene-flow [32,88]. Outbreeding depression is considered to be less of a risk [31,32] as it is most likely when combining lineages with greater genetic distinction than we observed here [32]. We consider the greatest risk for *S. napiformis* to be associated with losing overall genetic variation if entire sites/regions are extirpated [27].

On the basis of our observation of sub-optimal gene-flow, we recommend the establishment of intermediate sites, with a particular focus on the southwest region, which we identified as the most likely to persist beyond our 50-year projection. As introduction/translocation only impacts genetic structure or adaptation when migrants establish and reproduce, the choice of recipient sites and ongoing monitoring is crucial to maximise the long-term success of any conservation efforts. Patterns of genetic diversity are not necessarily related to adaptive traits; therefore, quantitative approaches are likely to improve conservation success (see [6,7]).

Wide road verges (including some travelling stock routes (TSR) in New South Wales) often consist of high-quality native grassland/open woodland vegetation and some contain remnant *S. napiformis* populations. Roadside areas where *S. napiformis* is currently absent should be considered as candidate sites for establishment of corridors to facilitate gene-flow, ideally comprising plants with mixed genetics from each of the three regions. The provision of dispersal corridors by maximising the use of substantial TSRs and road verges in the area is a feasible and cost-effective conservation action. However, we believe the most beneficial conservation measure would be to open converted land for restoration of native grasslands, which would have wider benefits for the entire southeast Australian grassland system and the organisms that rely on it.

Supplementary Materials: The following are available online at http://www.mdpi.com/1424-2818/12/11/417/s1, Figure S1: All regions. DAPC: Bayesian information criterion (BIC) and assignments; fastStructure: DeltaK/ProbK. Figure S2: Central + northeast. DAPC: BIC, DAPC plot, assignments; fastStructure: DeltaK/ProbK. Figure S3: Southwest. DAPC: BIC, DAPC plot, assignments; fastStructure: DeltaK/ProbK. Figure S4: Haplotype network and geographic distribution of haplotypes based on mapping of ddRADseq loci to chloroplast reference genomes. Figure S5: Percentage contributions of soil and climate variables when modelling the distribution of e on the basis of occurrence data. Table S1: List of variables used for modelling suitable habitat for *Sclerolaena napiformis*. Table S2: Analysis of molecular variance among disjunct southwest, central, and northeast regions of *Sclerolaena napiformis*. Table S3: Analysis of molecular variance among southwestern *Sclerolaena napiformis* sites. Table S4: Analysis of molecular variance among central *Sclerolaena napiformis* sites. Analysis includes investigations of variance (a) between Victoria and New South Wales (states are divided by the Murray River), and (b) among all sites without consideration of state variable. Table S5: Analysis of molecular variance among northeast *Sclerolaena napiformis* sites. Table S6: Lower, mean and upper FIS estimates for each collection site of *Sclerolaena napiformis*.

Author Contributions: Conceptualisation, M.D.A., E.A.J. and N.G.W.; methodology, M.D.A., E.A.J., and N.G.W.; formal analysis, M.D.A.; investigation, M.D.A., E.A.J., and N.G.W.; resources, E.A.J. and N.G.W.; data curation, N.G.W. and M.D.A.; writing—original draft preparation, M.D.A. and E.A.J.; writing—review and editing, M.D.A., E.A.J., and N.G.W.; visualisation, M.D.A. and E.J.; supervision, E.A.J.; project administration, E.A.J.; funding acquisition, E.A.J. and N.G.W. All authors have read and agreed to the published version of the manuscript.

Funding: This research was funded by Department of the Environment and Energy, Australian Government, grant number TSRF-161 and the Royal Botanic Gardens Victoria (RBGV). The article processing charge was funded by the RBGV.

Acknowledgments: Gareth Holmes (RBGV) for preparing DNA libraries for genome sequencing; Chris Jackson (RBGV) for advice and assistance with data handling; Mirinda Thorpe, Iestyn Hosking, and Karly Learmonth (Victorian Department of Environment Land Water and Planning) for locality information, landholders, and local stakeholders who provided information, transport, and arranged access to sites; David Hines and Deanna Marshall (Trust for Nature) for incorporating plants from our collection into conservation plantings on private landholdings on the Patho Plains; and Hayley Cameron (Monash University) for valuable input into an earlier version of the manuscript.

Conflicts of Interest: The authors declare no conflict of interest. The funders had no role in the design of the study; in the collection, analyses, or interpretation of data; in the writing of the manuscript; or in the decision to publish the results.

Data Accessibility: Raw sequence data, assemblies and associated scripts can be accessed via Mendeley data (http://dx.doi.org/10.17632/664sh75jgz.1).

References

1.	International Union for Conservation of Nature (IUCN). The IUCN Red List of Threatened Species. 2020. Available online: https://www.iucnredlist.org (accessed on 21 May 2020).

2.	Garner, B.A.; Hoban, S.; Luikart, G. IUCN red list and the value of integrating genetics. *Conserv. Genet.* **2020**, *21*, 795–801. [CrossRef]

3.	McGuigan, K.; Sgrò, C.M. Evolutionary consequences of cryptic genetic variation. *Trends Ecol. Evol.* **2009**, *24*, 305–311. [CrossRef] [PubMed]

4.	Sgrò, C.M.; Lowe, A.J.; Hoffmann, A.A. Building evolutionary resilience for conserving biodiversity under climate change. *Evol. Appl.* **2011**, *4*, 326–337. [CrossRef] [PubMed]

5.	Hoffmann, A.; Griffin, P.; Dillon, S.; Catullo, R.; Rane, R.; Byrne, M.; Jordan, R.; Oakeshott, J.; Weeks, A.; Joseph, L.; et al. A framework for incorporating evolutionary genomics into biodiversity conservation and management. *Clim. Chang. Responses* **2015**, *2*, 1–23. [CrossRef]

6.	Reed, D.H.; Frankham, R. Correlation between fitness and genetic diversity. *Conserv. Biol.* **2003**, *17*, 230–237. [CrossRef]

7.	Kramer, A.T.; Havens, K. Plant conservation genetics in a changing world. *Trends Plant Sci.* **2009**, *14*, 599–607. [CrossRef] [PubMed]

8.	Ouborg, N.J.; Vergeer, P.; Mix, C. The rough edges of the conservation genetics paradigm for plants. *J. Ecol.* **2006**, *94*, 1233–1248. [CrossRef]

9.	Dunlop, M.; Hilbert, D.; Ferrier, S.; House, A.; Liedloff, A.; Prober, S.M.; Smyth, A.; Martin, T.G.; Harwood, T.; Williams, K.J.; et al. A report prepared for the Department of Sustainability, Environment, Water, Population and Communities and the Department of Climate Change and Energy Efficiency. In *The Implications of Climate Change for Biodiversity Conservation and the National Reserve System: Final Synthesis*; CSIRO Climate Adaptation Flagship: Canberra, Australia, 2012.

10. Mable, B.K. Conservation of adaptive potential and functional diversity: Integrating old and new approaches. *Conserv. Genet.* **2019**, *20*, 89–100. [CrossRef]

11. Pina-Martins, F.; Baptista, J.; Pappas, G., Jr.; Paulo, O.S. New insights into adaptation and population structure of cork oak using genotyping by sequencing. *Glob. Chang. Biol.* **2019**, *25*, 337–350. [CrossRef]

12. Willi, Y.; Hoffmann, A.A. Demographic factors and genetic variation influence population persistence under environmental change. *J. Evol. Biol.* **2009**, *22*, 124–133. [CrossRef]

13. Mimura, M.; Yahara, T.; Faith, D.P.; Vazquez-Dominguez, E.; Colautti, R.I.; Araki, H.; Javadi, F.; Nunez-Farfan, J.; Mori, A.S.; Zhou, S.; et al. Understanding and monitoring the consequences of human impacts on intraspecific variation. *Evol. Appl.* **2017**, *10*, 121–139. [CrossRef] [PubMed]

14. Willi, Y.; Van Buskirk, J.; Hoffmann, A.A. Limits to the adaptive potential of small populations. *Annu. Rev. Ecol. Evol. Syst.* **2006**, *37*, 433–458. [CrossRef]

15. Vilas, A.; Perez-Figueroa, A.; Quesada, H.; Caballero, A. Allelic diversity for neutral markers retains a higher adaptive potential for quantitative traits than expected heterozygosity. *Mol. Ecol.* **2015**, *24*, 4419–4432. [CrossRef] [PubMed]

16. Hoffmann, A.A.; Sgrò, C.M. Climate change and evolutionary adaptation. *Nature* **2011**, *470*, 479–485. [CrossRef] [PubMed]

17. Cunze, S.; Heydel, F.; Tackenberg, O. Are plant species able to keep pace with the rapidly changing climate. *PLoS ONE* **2013**, *8*, e67909. [CrossRef]

18. Vitt, P.; Havens, K.; Kramer, A.T.; Sollenberger, D.; Yates, E. Assisted migration of plants: Changes in latitudes, changes in attitudes. *Biol. Conserv.* **2010**, *143*, 18–27. [CrossRef]

19. Franklin, I.R.; Frankham, R. How large must populations be to retain evolutionary potential? *Anim. Conserv.* **1998**, *1*, 69–70. [CrossRef]

20. Nunney, L.; Campbell, K.A. Assessing minimum viable population size–Demography meets population genetics. *Trends Ecol. Evol.* **1993**, *8*, 234–239. [CrossRef]

21. Commonwealth of Australia. Interim Biogeographical Regionalisation of Australia, Version 7. 2020. Available online: https://www.environment.gov.au/land/nrs/science/ibra (accessed on 17 July 2020).

22. EPBC. Environment Protection and Biodiversity Conservation Act 1999. 2016. Available online: https://www.legislation.gov.au/Details/C2016C00777 (accessed on 17 July 2020).

23. Mavromihalis, J. *National Recovery Plan for the Turnip Copperburr Sclerolaena napiformis*; Victorian Government Department of Sustainability and Environment (DSE): Melbourne, Australia, 2010.

24. Cunningham, G.M.; Mulham, W.E.; Milthorpe, P.L.; Leigh, J.H. *Plants of Western New South Wales*; CSIRO Publishing: Melbourne, Australia, 1992.

25. Conn, B.J. Natural regions and vegetation of Victoria. In *Flora of Victoria*; Foreman, D.B., Walsh, N.G., Eds.; Inkata Press: Melbourne, Australia, 1993; Volume 1, pp. 79–158.

26. Van Rossum, F.; Martin, H.; Le Cadre, S.; Brachi, B.; Christenhusz, M.J.M.; Touzet, P. Phylogeography of a widely distributed species reveals a cryptic assemblage of distinct genetic lineages needing separate conservation strategies. *Perspect. Plant. Ecol. Evol. Syst.* **2018**, *35*, 44–51. [CrossRef]

27. Weeks, A.R.; Sgrò, C.M.; Young, A.G.; Frankham, R.; Mitchell, N.J.; Miller, K.A.; Byrne, M.; Coates, D.J.; Eldridge, M.D.; Sunnucks, P.; et al. Assessing the benefits and risks of translocations in changing environments: A genetic perspective. *Evol. Appl.* **2011**, *4*, 709–725. [CrossRef]

28. Prober, S.M.; Byrne, M.; McLean, E.H.; Steane, D.A.; Potts, B.M.; Vaillancourt, R.E.; Stock, W.D. Climate-adjusted provenancing: A strategy for climate-resilient ecological restoration. *Front. Ecol. Evol.* **2015**, *3*, 65. [CrossRef]

29. Walsh, N.G. *Maireana obrienii* (Chenopodiaceae), a new species from eastern Australia. *Muelleria* **2013**, *31*, 61–64.

30. Walsh, N.G.; Sluiter, I.R.K. Lectotypification of *Atriplex stipitata* Benth. (Chenopodiaceae) and recognition of a new subspecies. *Muelleria* **2020**, *38*, 101–109.

31. Frankham, R. Genetic rescue of small inbred populations: Meta-analysis reveals large and consistent benefits of gene flow. *Mol. Ecol.* **2015**, *24*, 2610–2618. [CrossRef]

32. Frankham, R.; Ballou, J.D.; Eldridge, M.D.; Lacy, R.C.; Ralls, K.; Dudash, M.R.; Fenster, C.B. Predicting the probability of outbreeding depression. *Conserv. Biol.* **2011**, *25*, 465–475. [CrossRef]

33. Wilson, P.G. *Sclerolaena napiformis* Paul G. *Wilson sp. nov. Flora Aust.* **1984**, *4*, 330.

34. Angiosperm Phylogeny Group (APG). An update of the classification for the orders and families of flowering plants: APG IV. *Bot. J. Linn. Soc.* **2016**, *181*, 1–20. [CrossRef]

35. Kadereit, G.; Newton, R.J.; Vandelook, F. Evolutionary ecology of fast seed germination—A case study in Amaranthaceae/Chenopodiaceae. *Perspect. Plant. Ecol. Evol. Syst.* **2017**, *29*, 1–11. [CrossRef]

36. Woinarski, J.C.Z.; Braby, M.F.; Burbidge, A.A.; Coates, D.; Garnett, S.T.; Fensham, R.J.; Legge, S.M.; McKenzie, N.L.; Silcock, J.L.; Murphy, B.P. Reading the black book: The number, timing, distribution and causes of listed extinctions in Australia. *Biol. Conserv.* **2019**, *239*, 108261. [CrossRef]

37. Peterson, B.K.; Weber, J.N.; Kay, E.H.; Fisher, H.S.; Hoekstra, H.E. Double digest RADseq: An inexpensive method for de novo SNP discovery and genotyping in model and non-model species. *PLoS ONE* **2012**, *7*, 1–11. [CrossRef]

38. Amor, M.D.; Jackson, C.J.; Holmes, G.D.; James, E.A. Characterization of the complete chloroplast genome of *Sclerolaena napiformis* Wilson, an endangered Australian chenopod. *Mitochondrial DNA B* **2020**, *5*, 1332–1333. [CrossRef]

39. Murray, B.; Young, A.G. Widespread chromosome variation in the endangered grassland forb *Rutidosis leptorrhynchoides* F. Muell. (Asteraceae: Gnaphalieae). *Ann. Bot.* **2001**, *87*, 83–90. [CrossRef]

40. Jackson, R.C. Chromosomal evolution in *Haplopappus gracilis*: A centric transposition race. *Evolution* **1973**, *27*, 243–256.

41. Bolger, A.M.; Lohse, M.; Usadel, B. Trimmomatic: A flexible trimmer for Illumina Sequence Data. *Bioinformatics* **2014**, *30*, 2114–2120. [CrossRef]

42. Bankevich, A.; Nurk, S.; Antipov, D.E.A. SPAdes: A new genome assembly algorithm and its applications to single-cell sequencing. *J. Comput. Biol.* **2012**, *19*, 455–477. [CrossRef]

43. Seppey, M.; Manni, M.; Zdobnov, E.M. BUSCO: Assessing genome assembly and annotation completeness. In *Gene Prediction*; Methods in Molecular Biology Series; Kollmar, M., Ed.; Humana: New York, NY, USA, 2019; Volume 1962, pp. 227–245.

44. Wood, D.E.; Salzburg, S.L. Kraken: Ultrafast metagenomic sequence classification using exact alignments. *Genome Biol.* **2014**, *1415*, R46. [CrossRef]

45. Catchen, J.; Hohenlohe, P.; Bassham, S.; Amores, A.; Cresko, W. Stacks: An analysis tool set for population genomics. *Mol. Ecol.* **2013**, *22*, 3124–3140. [CrossRef]

46. Eaton, D.A.R.; Overcast, I. Ipyrad: Interactive assembly and analysis of RADseq datasets. *Bioinformatics* **2020**, *36*, 2592–2594. [CrossRef] [PubMed]

47. Foll, M.; Gaggiotti, O.E. A genome scan method to identify selected loci appropriate for both dominant and codominant markers: A Bayesian perspective. *Genetics* **2008**, *180*, 977–993. [CrossRef] [PubMed]

48. Luu, K.; Blum, M.; Privé, F. Pcadapt: Fast Principal Component Analysis for Outlier Detection. R package Version 4.1.0. 2019. Available online: https://CRAN.R-project.org/package=pcadapt (accessed on 4 July 2020).

49. R Core Team. *R: A Language and Environment for Statistical Computing*; R Foundation for Statistical Computing: Vienna, Austria, 2019; Available online: http://www.r-project.org (accessed on 1 June 2020).

50. Paradis, E. Pegas: An R package for population genetics with an integrated–modular approach. *Bioinformatics* **2010**, *26*, 419–420. [CrossRef]

51. Raj, A.; Stephens, M.; Pritchard, J.K. FastSTRUCTURE: Variational inference of population structure in large SNP data sets. *Genetics* **2014**, *197*, 573–589. [CrossRef]

52. Danecek, P.; Auton, A.; Abecasis, G.; Albers, C.A.; Banks, E.; DePristo, M.A.; Handsaker, R.E.; Lunter, G.; Marth, G.T.; Sherry, S.T.; et al. The variant call format and VCFtools. *Bioinformatics* **2011**, *27*, 2156–2158. [CrossRef] [PubMed]

53. Purcell, S.; Neale, B.; Todd-Brown, K.; Thomas, L.; Ferreira, M.A.R.; Bender, D.; Maller, J.; Sklar, P.; de Bakker, P.I.W.; Daly, M.J.; et al. PLINK: A toolset for whole-genome association and population-based linkage analysis. *Am. J. Hum. Genet.* **2007**, *81*, 559–575. [CrossRef]

54. Jombart, T.; Devillard, S.; Balloux, F. Discriminant analysis of principal components: A new method for the analysis of spatially structured populations. *BMC Genet.* **2010**, *11*, 94. [CrossRef]

55. Oksanen, J.; Blanchet, F.G.; Friendly, M.; Kindt, R.; Legendre, P.; McGlinn, D.; Minchin, P.R.; O'Hara, R.B.; Simpson, G.L.; Solymos, P.; et al. Vegan: Community Ecology Package. R package Version 2.5-4. 2019. Available online: https://cran.r-project.org/package=vegan (accessed on 4 July 2020).

56. Kamvar, Z.N.; Tabima, J.F.; Grunwald, N.J. Poppr: An R package for genetic analysis of populations with clonal, partially clonal, and/or sexual reproduction. *PeerJ* **2014**, *2*, e281. [CrossRef]

57. Goudet, J. Hierfstat, a package for R to compute and test hierarchical F-statistics. *Mol. Ecol. Notes* **2005**, *5*, 184–186. [CrossRef]

58. Winter, D.J. Mmod: An R library for the calculation of population differentiation statistics. *Mol. Ecol. Resour.* **2012**, *12*, 1158–1160. [CrossRef]

59. Hijmans, R.J.; Phillips, S.; Leathwick, J.; Elith, J. Package 'Dismo'. 2011. Available online: http://cran.r-project.org/web/packages/dismo/index.html (accessed on 4 July 2020).

60. Hijmans, R.J.; van Etten, J. Raster: Geographic Analysis and Modeling with Raster Data. R Package Version 2.0-12. 2012. Available online: http://CRAN.R-project.org/package=raster (accessed on 4 July 2020).

61. Auld, B.A.; Martin, P.M. Morphology and distribution of *Bassia birchii* (F. Muell.) F. Muell. [native sheep forage shrub, semi-arid eastern Australia]. *Proc. Linn. Soc. N. S. W.* **1976**, *100*, 167–178.

62. Novaes, R.M.; De Lemos Filho, J.P.; Ribeiro, R.A.; Lovato, M.B. Phylogeography of *Plathymenia reticulata* (Leguminosae) reveals patterns of recent range expansion towards northeastern Brazil and southern Cerrados in Eastern Tropical South America. *Mol. Ecol.* **2010**, *19*, 985–998. [CrossRef]

63. Peter, B.M.; Slatkin, M. Detecting range expansions from genetic data. *Evolution* **2013**, *67*, 3274–3289. [CrossRef]

64. Amor, M.D.; Johnson, J.C.; James, E.A. Identification of clonemates and genetic lineages using next-generation sequencing (ddRADseq) guides conservation of a rare species, *Bossiaea vombata* (Fabaceae). *Perspect. Plant Ecol. Evol. Syst.* **2020**, *45*, 125544. [CrossRef]

65. Harper, J.L. *Population Biology of Plants*; Academic Press: London, UK, 1977.

66. Silvertown, J.W.; Lovett-Doust, J. *Introduction to Plant Population Biology*; Blackwell: Oxford, UK, 1993.

67. Ouborg, N.J.; Piquot, Y.; Groenendael, V. Population genetics, molecular markers and the study of dispersal in plants. *J. Ecol.* **1999**, *87*, 551–568. [CrossRef]

68. James, E.A.; Jordan, R. Limited structure and widespread diversity suggest potential buffers to genetic erosion in a threatened grassland shrub *Pimelea spinescens* (Thymelaeaceae). *Conserv. Genet.* **2014**, *15*, 305–317. [CrossRef]

69. Llorens, T.M.; Ayre, D.J.; Whelan, R. Anthropogenic fragmentation may not alter pre-existing patterns of genetic diversity and differentiation in perennial shrubs. *Mol. Ecol.* **2018**, *27*, 1541–1555. [CrossRef] [PubMed]

70. Aguilar, R.; Quesada, M.; Ashworth, L.; Herrerias-Diego, Y.; Lobo, J. Genetic consequences of habitat fragmentation in plant populations: Susceptible signals in plant traits and methodological approaches. *Mol. Ecol.* **2008**, *17*, 5177–5188. [CrossRef]

71. McDougall, K.; Kirkpatrick, J.B. *Conservation of Lowland Native Grassland in South-Eastern Australia*; World Wide Fund for Nature: Sydney, Australia, 1994.

72. Commonwealth of Australia. *The National Recovery Plan for the Plains-Wanderer (Pedionomus torquatus)*; Department of the Environment: Canberra, Australia, 2016.

73. Peakall, R.; Oliver, I.; Turnbull, C.L.; Beattie, A.J. Genetic diversity in an ant-dispersed chenopod *Sclerolaena diacantha*. *Aust. J. Ecol.* **1993**, *18*, 171–179. [CrossRef]

74. Peakall, R.; Beattie, A.J. Does ant dispersal of seeds in *Sclerolaena diacantha* (Chenopodiaceae) generate local spatial genetic structure? *Heredity* **1995**, *75*, 351. [CrossRef]

75. Goodwillie, C.; Kalisz, S.; Eckert, C.G. The evolutionary enigma of mixed mating systems in plants: Occurrence, theoretical explanations, and empirical evidence. *Ann. Rev. Ecol. Evol. Syst.* **2005**, *36*, 47–79. [CrossRef]

76. Laenen, B.; Tedder, A.; Nowak, M.D.; Toräng, P.; Wunder, J.; Wötzel, S.; Steige, K.A.; Kourmpetis, Y.; Odong, T.; Drouzas, A.D.; et al. Demography and mating system shape the genomewide impact of purifying selection in *Arabis alpina*. *Proc. Natl. Acad. Sci. USA* **2018**, *115*, 816–821. [CrossRef]

77. Eckert, C.G.; Kalisz, S.; Geber, M.A.; Sargent, R.; Elle, E.; Cheptou, P.O.; Goodwillie, C.; Johnston, M.O.; Kelly, J.K.; Moeller, D.A.; et al. Plant mating systems in a changing world. *Trends Ecol. Evol.* **2010**, *25*, 35–43. [CrossRef]

78. Jordan, C.Y.; Lohse, K.; Turner, F.; Thomson, M.; Gharbi, K.; Ennos, R.A. Maintaining their genetic distance: Little evidence for introgression between widely hybridizing species of *Geum* with contrasting mating systems. *Mol. Ecol.* **2018**, *27*, 1214–1228. [CrossRef]

79. Mustin, K.; Benton, T.G.; Dytham, C.; Travis, J.M.J. The dynamics of climate-induced range shifting; Perspectives from simulation modelling. *Oikos* **2009**, *118*, 131–137. [CrossRef]

80. Scott, J.K.; Webber, B.L.; Murphy, H.; Kriticos, D.J.; Ota, N.; Loechel, B. Weeds and Climate Change: Supporting Weed Management Adaptation. 2014. Available online: www.AdaptNRM.org (accessed on 18 September 2020).
81. Jurado, E.; Westoby, M. Germination biology of selected central Australian plants. *Aust. J. Ecol.* **1992**, *17*, 341–348. [CrossRef]
82. Navie, S.C.; Cowley, R.A.; Rogers, R.W. The relationship between distance from water and the soil seed bank in a grazed semi-arid subtropical rangeland. *Aust. J. Bot.* **1996**, *44*, 421–431. [CrossRef]
83. Auld, B.A. Aspects of the population ecology of galvanised burr (*Sclerolaena birchii*). *Aust. Range. J.* **1981**, *3*, 142–148. [CrossRef]
84. Carta, F.E.; Parsons, R.F. Notes on the germination of the endangered species *Sclerolaena napiformis* (Chenopodiaceae). *Cunninghamiana* **2005**, *9*, 215–218.
85. Plummer, J.A.; Bell, D.T. The effect of temperature, light and gibberellic acid (GA3) on the germination of Australian everlasting daisies (Asteraceae, tribe Inuleae). *Aust. J. Bot.* **1995**, *43*, 93–102. [CrossRef]
86. Bykova, O.; Chuine, I.; Morin, X. Highlighting the importance of water availability in reproductive processes to understand climate change impacts on plant biodiversity. *Perspect. Plant. Ecol. Evol. Syst.* **2019**, *37*, 20–25. [CrossRef]
87. Kinloch, J.E.; Friedel, M.H. Soil seed reserves in arid grazing lands of central Australia. Part 1: Seed bank and vegetation dynamics. *J. Arid Environ.* **2005**, *60*, 133–161. [CrossRef]
88. Ralls, K.; Ballou, J.D.; Dudash, M.R.; Eldridge, M.D.B.; Fenster, C.B.; Lacy, R.C.; Sunnucks, P.; Frankham, R. Call for a paradigm shift in the genetic management of fragmented populations. *Conserv. Lett.* **2018**, *11*, e12412. [CrossRef]
89. Krauss, S.L.; Dixon, B.; Dixon, K.W. Rapid genetic decline in a translocated population of the endangered plant *Grevillea scapigera*. *Conserv. Biol.* **2002**, *16*, 986–994. [CrossRef]

Publisher's Note: MDPI stays neutral with regard to jurisdictional claims in published maps and institutional affiliations.

Article

Disentangling the Genetic Relationships of Three Closely Related Bandicoot Species across Southern and Western Australia

Rujiporn Thavornkanlapachai [1], Esther Levy [1], You Li [2], Steven J. B. Cooper [3,4], Margaret Byrne [1] and Kym Ottewell [1,*]

1   Department of Biodiversity, Conservation and Attractions, Locked Bag 104, Bentley Delivery Centre, Kensington, WA 6152, Australia; rujiporn.sun@dbca.wa.gov.au (R.T.); eslevy11@gmail.com (E.L.); margaret.byrne@dbca.wa.gov.au (M.B.)
2   School of Life Sciences and Engineering and Biomedical Research Centre, Northwest Minzu University, Lanzhou 730030, China; 284142549@xbmu.edu.cn
3   School of Biological Sciences, Australian Centre for Evolutionary Biology and Biodiversity, the University of Adelaide, Adelaide, SA 5005, Australia; Steve.Cooper@samuseum.sa.gov.au
4   Evolutionary Biology Unit, South Australian Museum, North Terrance, Adelaide, SA 5000, Australia
*   Correspondence: kym.ottewell@dbca.wa.gov.au

Abstract: The taxonomy of Australian *Isoodon* bandicoots has changed continuously over the last 20 years, with recent genetic studies indicating discordance of phylogeographic units with current taxonomic boundaries. Uncertainty over species relationships within southern and western *Isoodon*, encompassing *I. obesulus*, *I. auratus*, and *I. fusciventer*, has been ongoing and hampered by limited sampling in studies to date. Identification of taxonomic units remains a high priority, as all are threatened to varying extents by ongoing habitat loss and feral predation. To aid diagnosis of conservation units, we increased representative sampling of *I. auratus* and *I. fusciventer* from Western Australia (WA) and investigated genetic relationships of these with *I. obesulus* from South Australia (SA) and Victoria (Vic) using microsatellite markers and mitochondrial DNA. mtDNA analysis identified three major clades concordant with *I. obesulus* (Vic), *I. auratus*, and *I. fusciventer*; however, *I. obesulus* from SA was polyphyletic to WA taxa, complicating taxonomic inference. Microsatellite data aided identification of evolutionarily significant units consistent with existing taxonomy, with the exception of SA *I. obesulus*. Further, analyses indicated SA and Vic *I. obesulus* have low diversity, and these populations may require more conservation efforts than others to reduce further loss of genetic diversity.

Keywords: golden bandicoot; southern brown bandicoot; quenda; phylogenetic; taxonomy; evolutionarily significant unit; ecosystem engineer

Citation: Thavornkanlapachai, R.; Levy, E.; Li, Y.; Cooper, S.J.B.; Byrne, M.; Ottewell, K. Disentangling the Genetic Relationships of Three Closely Related Bandicoot Species across Southern and Western Australia. *Diversity* 2021, 13, 2. https://dx.doi.org/10.3390/d13010002

Received: 23 November 2020
Accepted: 7 December 2020
Published: 22 December 2020

Publisher's Note: MDPI stays neutral with regard to jurisdictional claims in published maps and institutional affiliations.

1. Introduction

Threatened species often require active management to ensure their persistence. An important first step in threatened species conservation is identifying and defining appropriate units of management to direct conservation effort to where it is needed the most [1,2]. Confusion about taxonomic boundaries of subspecies and other conservation units has the potential to lead to inappropriate management decisions that may have detrimental consequences. For example, mixing populations that are genetically and evolutionarily distinct could result in the loss of local adaptation or outbreeding depression [3,4]. Conversely, managing populations as separate units, particularly when population sizes are small, may lead to increased levels of inbreeding and loss of adaptive potential due to random genetic drift [5]. Developing an appropriate taxonomic classification that reflects the actual nature of species or subspecies is therefore essential to guide management decisions to improve conservation outcomes.

The concept of "evolutionarily significant units" (ESU) has been used to aid management of populations in recent decades. The concept was introduced by Ryder [6] to resolve "subspecies" into manageable units where an ESU is defined as a group of organisms with high genetic and ecological distinctiveness. With increasing use of molecular tools, Moritz [7] proposed a genetic criterion for defining an ESU as a group of organisms that are reciprocally monophyletic for mtDNA haplotypes and have diverged allele frequencies at nuclear loci, a definition that places emphasis on historical isolation of populations. A sub-category under an ESU is "management unit" (MU), which recognises populations that do not show reciprocal monophyly for mtDNA haplotypes but have diverged haplotype and allele frequencies at mitochondrial or nuclear loci to accommodate more contemporary population structuring [7]. The use of reciprocal monophyly as a criterion for defining an ESU, whilst popular, has been criticised as being too stringent [8,9]. Fraser and Bernatchez's [8] concept of "adaptive evolutionary conservation" offers a more flexible approach, which defines an ESU as "a lineage demonstrating highly restricted gene flow from other such lineages within the higher organization level (or lineage) of the species". These lineages are sufficiently isolated that each lineage has limited or no impact on the evolution, genetic variance, and demography of other lineages [8]. This approach integrates multiple ESU concepts, including Moritz's, to fulfil conservation goals, which under differing circumstances will warrant the use of some concepts over others.

The short-nosed bandicoots, genus *Isoodon* (family Peramelidae), is a small-to-medium-sized Australian marsupial that can be found across Australia. Bandicoots are terrestrial, omnivorous marsupials with a bounding gait and forelimbs adapted for digging for food [10]. As digging mammals, they play an important role as ecosystem engineers, contributing to altered soil nutrient and moisture dynamics that enhance plant germination and growth [11]. Their "rat-like" appearance and cryptic nature make them less attractive for conservation efforts. However, they are known for their complex biology of adaptive traits and evolutionary diversity [11,12]. These traits enable some *Isoodon* species to adapt to urban environments [13,14]. However, facing pressure from introduced predators and human disturbances, many *Isoodon* species have dramatically reduced in distribution since European settlement and persist in remnants of their formal ranges.

The taxonomy of the genus *Isoodon* has changed multiple times over the last 70 years and is currently still in flux. Up to eleven different species of *Isoodon* have been recognised at various times [15]; however, only three are formally recognised currently: *I. macrourus*, *I. obesulus*, and *I. auratus* [16,17], with additional taxa *I. peninsulae* [18] and *I. fusciventer* [19] recently proposed, both raised from subspecies (of *I. obesulus*) to full species rank. Most recently, comprehensive geographic and taxonomic sampling in a phylogeographic study of *I. obesulus* and *I. auratus* has indicated further complexity in the relationships of species and subspecies within these taxa [12]. One major finding was that the geographic distribution of *I. o. obesulus* is significantly more restricted than previously recognised, with specimens of *I. o. obesulus* from South Australia (SA; encompassing Mount Lofty Ranges, Fleurieu Peninsula, and Kangaroo Island) genetically differentiated from *I. o. obesulus* from south-eastern Australia and more closely allied with *I. fusciventer* from Western Australia (WA; [12,20]). Further, consistent with early molecular studies [15,21], Cooper et al. [12] failed to resolve a clear pattern of reciprocal monophyly between *I. fusciventer* (southern WA) and *I. auratus* (northern WA) in mitochondrial DNA, evidence that has previously been used to suggest the species be synonymised [15,21]. This suggestion based on genetic data contrasts with morphological data indicating that *I. obesulus* and *I. auratus* can be readily distinguished by a range of characters, including size, skull and teeth characters, and fur colour [15,19,22,23]. While monophyly could not be resolved, Cooper et al. [12] identified a lack of mitochondrial DNA haplotype sharing between the taxa as well as fixed allozyme differences and private nuclear gene haplotypes, suggesting a putative pattern of paraphyly of these taxa.

Here, we focus on identifying conservation units within the "southern and western" group of *Isoodon* bandicoots, WA taxa *Isoodon auratus*, *I. a. barrowensis* and *I. fusciventer*, and southern Australian taxa *I. obesulus* (Vic and SA), all of which are of conservation concern. *Isoodon* species have dramatically declined in their abundance and distribution in the last 220 years since European settlement of Australia [24,25], primarily as a result of introduced predators such as feral cats, foxes and black rats, altered fire regimes, disease, and habitat loss from clearing of native vegetation and habitat modification [26–28]. *Isoodon obesulus*, *I. fusciventer*, and *I. auratus* are listed as Near Threatened, Least Concern, and Near Threatened, respectively (International Union for Conservation of Nature red list/Environment Protection and Biodiversity Conservation (EPBC) Act, Australia). The subspecies *I. o. obesulus* in eastern and south-eastern Australia is listed as Endangered, although re-circumscription as suggested by analyses in Cooper et al. [12] means this subspecies is now more geographically restricted requiring re-evaluation of its conservation status; whilst other subspecies, *I. a. auratus* and *I. a. barrowensis*, are listed as Vulnerable under the EPBC act. These species/subspecies are currently subjected to different levels of conservation action to mitigate threats [26], with *I. a. auratus* and *I. o. obesulus* particularly targeted for translocations and reintroductions to island and mainland feral predator-free areas to increase their population size [29]. Clarification of taxa and conservation units is required to support conservation assessment and listing across the group and to ensure future conservation management activities are appropriately applied.

In this study, we expand sampling of representative *I. auratus* from northern WA and *I. fusciventer* from southern WA, as well as introduce additional molecular data from hyper-variable nuclear loci (microsatellites), to provide further resolution to the relationships amongst southern and western *Isoodon* bandicoots as raised in Cooper et al. [12]. Incorporating existing molecular data from Cooper et al. [12] and Li et al. [20] with our expanded sampling, we use a combination of phylogenetic and hierarchical population genetic approaches to identify putative ESUs across the southern and western *Isoodon* bandicoots, *I. auratus*, *I. fusciventer*, *I. obesulus* SA, and *I. o. obesulus* Vic. Within each regional population, we investigate geographical sub-structuring and assess patterns of mitochondrial and microsatellite genetic diversity to inform priorities for conservation genetic management.

2. Materials and Methods

2.1. Sample Collection, Microsatellite Genotyping, and Mitochondrial DNA Sequencing

We collated microsatellite and mitochondrial DNA (mtDNA) profiles from previously published sources with new sampling for a total of 731 and 218 additional samples, respectively. *Isoodon obesulus*, *I. fusciventer*, and *I. auratus* samples were distributed across multiple locations in Western Australia, South Australia, and Victoria and across multiple years from 2002 to 2018 (Figure 1, Table S1). Microsatellite profiles for WA specimens of *I. auratus* and *I. fusciventer* (*n* = 172) were predominantly obtained from previously published datasets in Ottewell et al. [30] and Ottewell et al. [13], respectively, with some additional new sampling from southwest WA (*I. fusciventer*) and the Kimberley region (*I. a. auratus*) (Table S1). Ear biopsy samples were opportunistically obtained by environmental consultants during urban fauna relocations and during population monitoring by the Australian Wildlife Conservancy (Kimberley samples). A small number of samples were obtained from roadkill animals by citizen scientistsunder the Department of Biodiversity, Conservation and Attractions license to use animals for scientific purposes (permit no. U10/2018). DNA extraction and microsatellite genotyping were conducted as described in Ottewell et al. [30] using 10 microsatellite primers (B3-2, B20-5, B34-2, Ioo2, Ioo4, Ioo6, Ioo7, Ioo8, Ioo10, and Ioo16). Briefly, amplification reactions were carried out using the Qiagen Multiplex Kit, following the manufacturer's instruction (Qiagen, Hilden, Germany). PCR products were separated on an Applied Biosystems 3130xl capillary sequencer (Foster City, CA, USA). Fragment sizes were determined with GeneScan-500 LIZ size standard (Applied Biosystems) and scored in Genemapper v5.0 software (Applied Biosystems). Microsatellite profiles from published datasets were re-scored with new samples to ensure consistency in allele scoring. Microsatellite genotypes from SA and Vic were obtained from Li et al.'s [20] study (*n* = 559). Due to the addition of M13 tags to primers used in this study, which added 30 bp to tandem repeat sizes, we re-amplified a subset of 15 samples (*n* = 5 from each of Mount Burr, Mount Lofty Ranges, Grampians populations) in our WA laboratory to calibrate genotype scores between SA/Vic and WA datasets.

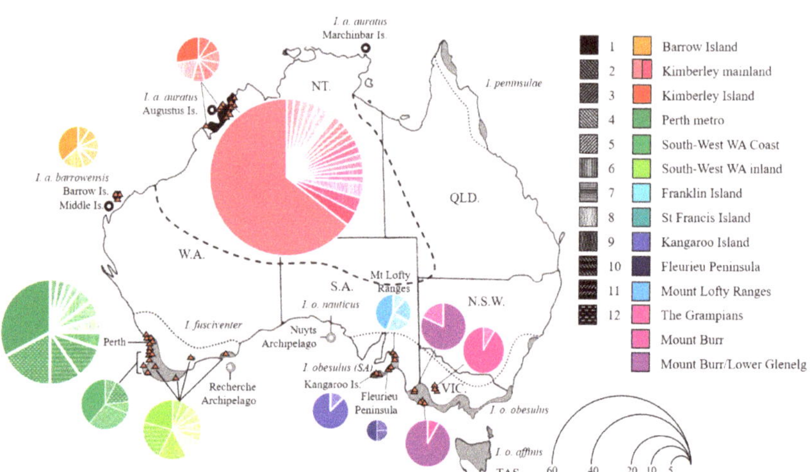

Figure 1. Map of current and historical distribution of *Isoodon* taxa included in the current study (adapted from Zenger et al. [21]) and geographic distribution of mitochondrial DNA haplotypes. Grey shading represents contemporary distribution of taxa formerly or currently recognised within *I. obesulus* and black shading represents current distribution of *I. auratus*. Dotted line indicates historical limits of *I. obesulus* and dashed lines designate historical *I. auratus* limits. ▲ indicates sampling locations. Each pie chart represents haplotype frequencies within each sampling site, and size of pie chart represents numbers of haplotypes. Colour codes indicate the sites where haplotypes were mostly found. Pattern codes represent individual haplotypes found within each site.

Sequences from the highly variable non-coding mitochondrial control region (D-loop) were obtained from Li et al. [20] (*n* = 49) and Cooper et al. [12] (*n* = 23, Table S1) with additional sequences obtained for a subset of the WA samples used in microsatellite genotyping (*n* = 146). New specimens were sequenced using primers L15999M and H16498M [31] with amplifications carried out in 25 µL reactions, which included ~10–20 ng of template DNA, 0.5 µM of each primer, 1.25 mM MgCl₂, 0.2 µM of each dNTPs, 0.05 µL of 10% BSA, 1× Reaction Buffer, and one unit of Invitrogen Taq polymerase. Reactions were run under the following conditions: 94 °C for 3 min followed by 35 cycles of denaturation 94 °C for 30 s, annealing at 55 °C for 45 s, and extension at 72 °C for 60 s, with a final extension of 72 °C for 7 min. Products were visualised on 1% agarose gels and Sanger-sequenced at a commercial service (Australian Genome Research Facility). Sequences were edited and aligned using SEQUENCHER v5.2 (Gene Codes Corporation, Ann Arbor, MI, USA) with consensus sequences from each of the datasets then aligned using CLUSTAL W with default parameters [32] and sequences trimmed to the same length in MEGA v10.0.5 [33].

2.2. Data Analysis

2.2.1. Phylogenetic Analysis

Phylogenetic analyses of mtDNA sequences were carried out using maximum likelihood in RAxML v7.0.3 [34], neighbour-joining in MEGA v10.0.5 [33] and Bayesian inference in BEAST v1.10.4 [35]. In RAxML, we applied a single model of evolution, General Time Reversible (GTR) model [36] with unequal variation at sites modelled using a Gamma (G) distribution [37] to the sequence data. Rapid bootstrap analysis was applied to search for the best-scoring maximum likelihood (ML) tree. In MEGA, a neighbour-joining tree was constructed using the proportion (*p*) of nucleotide site changes as a distance estimate and a Gamma (G) distribution as the rate among sites. The robustness of the nodes in the tree was assessed by 1000 bootstrap replicates. For BEAST analysis, we ran JMODEL-TEST v2.1.10 [38] to identify the most likely substitution model for our dataset. In BEAST, we used HKY+I+G model as inferred by the previous step with an uncorrelated lognormal relaxed clock. Four independent runs of 50 million generations and sampling every 5000 generations were carried out. The log files were analysed in TRACER v1.7.1 [39] to ensure stationarity had been reached by visually inspecting the traces and each log file had the effective sample sizes >200 for all parameters. Tree files were combined in LOGCOM-BINER v1.10.4 (BEAST package) with the burn-in period set to 2500 applied to each tree. The combined tree file was annotated using TREEANNOTATOR v1.10.4 (BEAST package). ML, neighbour-joining, and Bayesian trees were visualised using functions from the R packages "APE" [40] and "PHYTOOLS" [41] in R version 3.6.2 [42] using *I. macrourus* as an outgroup.

2.2.2. Population Level Analysis

To investigate genetic relationships between *Isoodon* from different clades, principal component analysis was carried out for both nuclear and mitochondrial DNA using the "ADE4" package [43] and the "ADEGENET" package [44] in R based on Euclidean genetic distances using the *dudi.pco* command. We did not scale allele or haplotype frequencies and left missing genotypes as is. The first two principal components were retained.

Multiple methods were used to infer population structure between and within major geographic regions (WA, SA, Vic) from nuclear DNA. Firstly, we used a Bayesian clustering in STRUCTURE 2.3.4 [45] based on a model that assumed admixture of ancestry and correlated allele frequencies. In each analysis, individuals were assigned a membership co-efficient, which is the fraction of the genome with membership to a particular genetic cluster. Ten independent runs were performed using 500,000 iterations, with a burn-in period of 50,000 iterations. We compared the likelihood values for different K values (1–20) and selected K based on the largest decrease in Delta K value [46,47]. Cluster membership coefficients were permuted over multiple runs using the Greedy algorithm in CLUMPP to obtain a mean across replicates [48]. Secondly, we used spatially explicit Bayesian clustering implemented in TESS 2.3.1 [49,50]. TESS differs from STRUCTURE in that it seeks genetic discontinuities in continuous populations and estimates individual admixture proportions using spatial prior information [49]. We used the admixture model with 50,000 sweeps and 10,000 burn-in for K ranging from 2–20 and 10 replicates per K. The optimal K was chosen where the value of the deviance information criterion (DIC) stabilises. The mean estimated cluster membership coefficient of the chosen K in TESS was obtained using CLUMPP as previously. Thirdly, unlike STRUCTURE and TESS, discriminant analysis of principal components (DAPC) does not assume Hardy–Weinberg equilibrium of clusters. DAPC is a multivariate method that partitions genetic variation into principal components to find groups using K-means that minimize within group variation [51]. We ran the *find.cluster* command using the ADEGENET package in R. We used the number of components that allowed 90% of cumulative variance to be retained. We selected the optimal K based on a combination of K with the lowest Bayesian information criterion (BIC) and after the largest decrease in BIC [51]. In resolving population structure, we undertook a hierarchical approach, firstly analysing all geographic regions together to identify the broadest genetic structuring, then successively removing the most distinctive cluster and repeating analyses to identify sub-structuring and hence putative conservation units down to the local population scale. Note that for regional analyses, we included the Mt Burr population, located in the far south-eastern South Australia (Figure 1), with other Victorian populations, as this population was shown to cluster with *I. o. obesulus* Vic [12,20].

Analysis of molecular variance (AMOVA) of mitochondrial DNA was applied in ARLEQUIN version 3.0 [52] to identify the distribution of genetic variation between and within hierarchical clusters identified by STRUCTURE. The locus-by-locus variance component was estimated from 16,000 permutations. Genetic divergences between pairs of population samples were quantified for nuclear DNA using Jost's [53] estimate of differentiation (D_{est}) using GENALEX version 6.5 [54]. While pairwise ϕ_{ST} values for mitochondrial DNA were quantified using ARLEQUIN version 3.0 with 16,000 permutations [52].

To test whether the observed haplotype patterns within *Isoodon* spp. populations indicate recent bottlenecks or, conversely, demographic expansion, we used Tajima's D and Fu's F_S neutrality tests and mismatch distribution carried out in ARLEQUIN version 3.0 [52]. In general, Tajima's D and Fu's F_S neutrality tests were used to assess whether the data were consistent with the population being at neutral mutation-drift equilibrium [55,56]. Departure of Tajima's D or Fu's F_S from zero, either negative or positive, indicates that the population of interest is deviated from the assumption of neutrality and/or equilibrium. If D or $F_S < 0$, the population size may be increasing or an indication of purifying selection. If D or $F_S > 0$, the population size may be decreasing (or bottleneck) or an indication of overdominant selection. We used 1000 simulated samples to calculate the level of significance for both tests. Mismatch distribution employs the distribution of pairwise differences between haplotypes from which spatial population expansion can be estimated. Multi-modal mismatch distribution testifies the population's demographic equilibrium or subdivided populations, while uni-modal suggests recent demographic expansion [57,58], or range expansions with a high level of gene flow [59,60]. We used 1000 bootstrap pseudoreplicates to test the model of spatial expansion by comparing the sum of squared deviations (SSD) between observed and simulated data. The Harpending's raggedness index (r) was also used to test for deviation from unimodality. Smaller values indicate a better fit. The significance of the estimate was also obtained from the corresponding p value.

2.2.3. Genetic Variation Analysis

We assessed the genotypic error rate of microsatellite scores between laboratories by calculating the allele- and locus-specific genotypic error rates [61]. We then tested for Hardy–Weinberg equilibrium (HWE) among loci for each population with sample size ≥ 10 using GENALEX version 6.51 [54]. We tested for the presence of null alleles at each locus using MICROCHECKER [62]. Estimates of the allelic richness (an estimate of the number of alleles per locus corrected for sample size), number of alleles, gene diversity, inbreeding coefficient (F_{IS}) were calculated using FSTAT version 2.9.3.2 [63]. The significant deviation of F_{IS} values from Hardy–Weinberg Equilibrium was determined by randomization tests in FSTAT. Differences in gene diversity and allelic richness among regions were statistically tested using a non-parametric Friedman's test with locus as a blocking factor and post-hoc multiple comparisons made using the Conover test with Bonferroni correction in the R package "PMCMR" [64]. Mitochondrial DNA diversity was quantified by calculating the number of haplotypes, gene diversity, and nucleotide diversity in ARLEQUIN version 3.0 [52].

3. Results

3.1. Genetic Data Quality Assessment

Across 218 mtDNA sequenced samples, 96 polymorphic sites, including sites with gaps, were found in the 497 bp D-loop region, giving a total of 77 haplotypes. Across the microsatellite dataset, the amplification success rate was 0.926 per locus. The allele-specific genotypic error rate from 15 re-genotyped samples was 0.107 (0.030 due to allelic drop-out and 0.077 due to false alleles). All loci deviated from HWE at least in one location, but none consistently deviated from HWE across all locations. Likewise, MICROCHECKER detected putative null alleles in all loci, but they were not consistent in all locations. All loci were retained for further analysis.

3.2. Phylogeographic Structure across Southern and Western Australia

Phylogenetic analysis of mtDNA using Bayesian methods resolved three major evolutionary groups within our sample (Figure 2). At the highest level, *I. o. obesulus* from Victoria was distinguished from the remaining SA and WA entities with very high support (PP 1.0; Figure 2). Additional sampling helped us to further resolve two evolutionary groups within the latter clade of southern and western *Isoodon*, with northern WA taxa (*I. a. auratus*, *I. a. barrowensis*), for the most part, distinct from southern WA (*I. fusciventer*, PP 0.93). However, mtDNA haplotypes from SA *I. obesulus* were polyphyletic, occurring across both clades. While *I. obesulus* on Kangaroo Island and Fleurieu Peninsula clustered with southwest WA *I. fusciventer*, *I. obesulus* in Mount Lofty Ranges was paraphyletic in both *I. fusciventer* and *I. auratus* clades, although with low internal posterior support (PP < 0.90, Figure 2). Similarly, a small number of *I. auratus* and *I. fusciventer* haplotypes were placed within opposing clades, again with low support (PP < 0.90, Figure 2). Maximum likelihood and neighbour-joining trees showed a similar pattern of phylogenetic clustering as the Bayesian tree with three major evolutionary groups formed, however, with differing relationships amongst clades (Figures S1 and S2). In both ML and NJ trees, SA *I. obesulus* were polyphyletic across *I. fusciventer* and *I. auratus* in a similar pattern to the Bayesian tree.

Nuclear (microsatellite) markers further confirmed support for the distinction of Mount Lofty Range, Fleurieu Peninsula, and Kangaroo Island *I. obesulus* in South Australia from Mount Burr in south-east SA and Victorian *I. obesulus*, with two main clusters (Mount Burr/Victoria *I. obesulus* and SA/WA *Isoodon*) evident in the PCoA (Figure 3a). This latter cluster indicated a geographic cline and genetic overlap from South Australia (*I. obesulus*), southern Western Australia (*I. fusciventer*) to northern Western Australia (*I. auratus*) (Figure 3b), which was similarly supported by PCoA analysis of mtDNA (Figure 3c,d).

At the highest level, all three clustering analyses (Structure, TESS, and DAPC) resolved K = 3 genetic clusters supporting the clear distinction of Mount Burr and Victorian *I. o. obesulus* from SA/WA (Figure 4a and Figure S3), but which failed to resolve distinction between the SA and WA taxa as currently recognised (i.e., *I. auratus*, *I. fusciventer*, SA *I. obesulus*). SA and northern WA, the two most geographically distant populations, formed the basis of two additional genetic clusters with a pattern of admixture occurring in southern WA (Figure 4a). This admixture pattern was also retained when Victorian *I. o. obesulus* was removed from analyses (Figure 4b and Figure S4).

Structuring within WA taxa based on microsatellite data was only revealed upon hierarchical analysis. With SA samples removed, structure analysis recovered K = 3 populations representing southern WA (*I. fusciventer*), Barrow Island (*I. a. barrowensis*), and the Kimberley (*I. a. auratus*) (Figure 4c). These clades are largely represented in phylogenetic analyses with subspecies *I. a. barrowensis* forming a distinct clade in maximum likelihood and neighbour-joining trees (Figures S1 and S2) but nested within a group comprising other Kimberley mainland samples in Bayesian analyses (Figure 2).

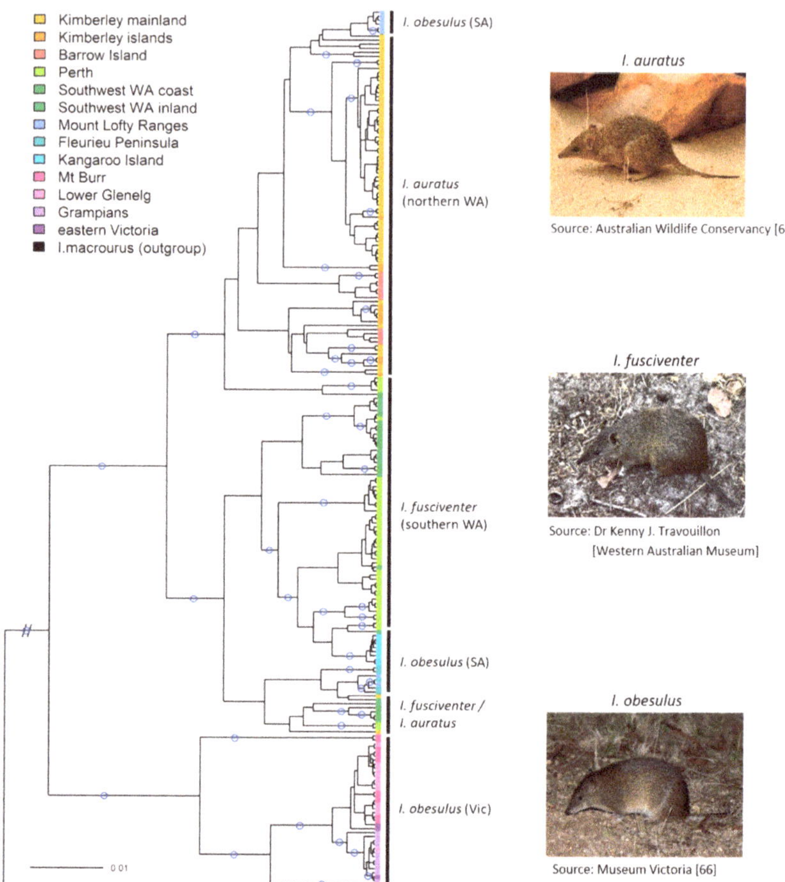

Figure 2. Results of Bayesian phylogenetic analysis in BEAST of D-loop mitochondrial sequences from Isoodon obesulus (Vic, SA), *I. fusciventer*, and *I. auratus*. Posterior probability values above 0.9 are indicated by grey circles overlaid on branches. *Isoodon macrourus* was used as an outgroup. Photo credits: *I. auratus* Australian Wildlife Conservancy [65], *I. fusciventer* Dr Kenny Travouillon, Western Australian Museum, *I. obesulus* Museums Victoria [66].

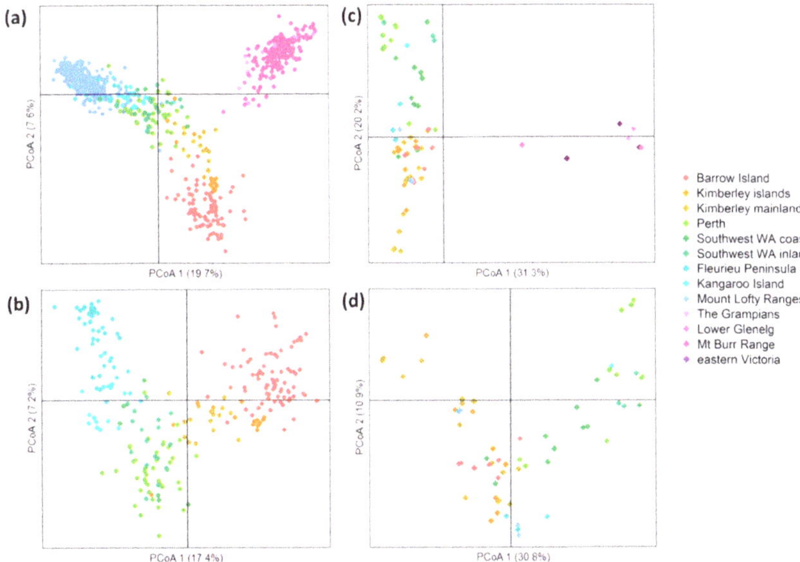

Figure 3. Stepwise principal co-ordinates analysis (PCoA) of allele frequencies based on ten microsatellite loci (**a**,**b**) and D-loop mtDNA haplotypes (**c**,**d**) including (**a**,**c**) and excluding (**b**,**d**) Victorian *I. obesulus* samples. Relative PCoA scores have been plotted for the first two dimensions (percentage variation explained on each axis). Individuals are represented in dots, and colours indicate sampling sites.

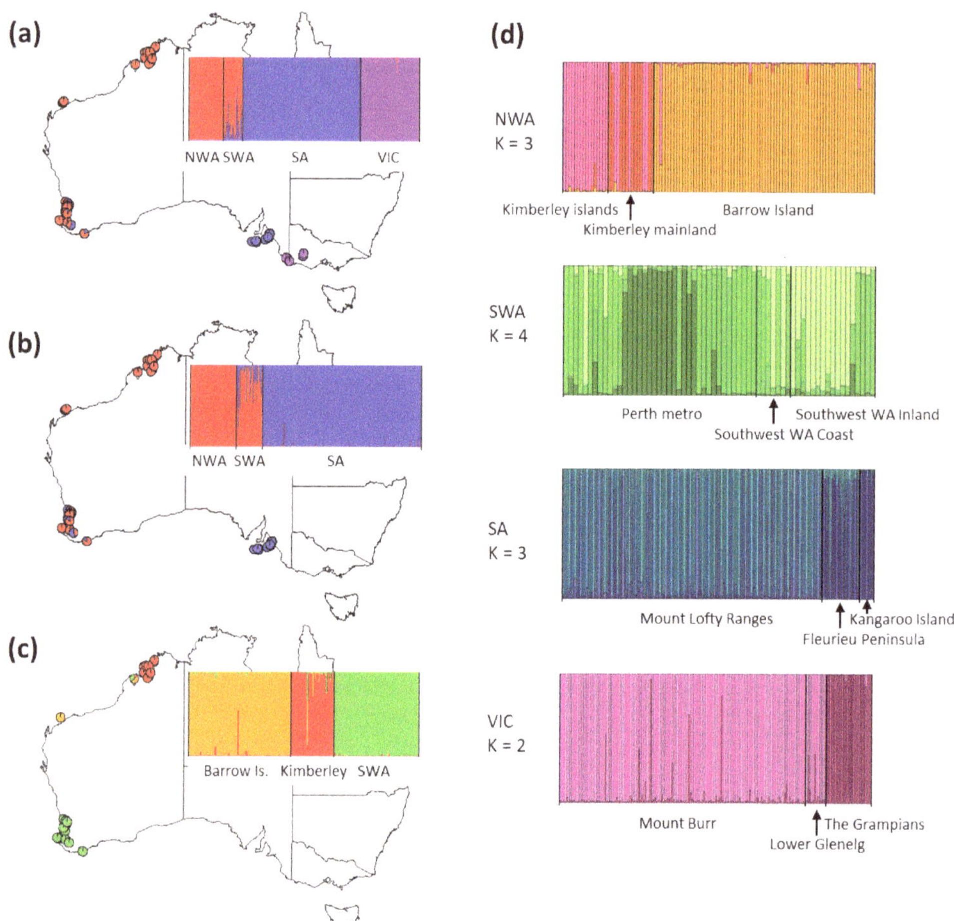

Figure 4. Summary of the hierarchical clustering analysis using STRUCTURE based on 10 microsatellite loci. The analyses include (**a**) all samples, (**b**) excluding Victoria, (**c**) excluding Victoria and South Australia, and (**d**) regional populations. Pie charts on Australia maps illustrate individuals' estimated membership to each genetic cluster relative to their geographic locations. Barplots in (**d**) show the most likely K value for each region. Regions include northern Western Australia (NWA), southern Western Australia (SWA), South Australia (SA), and Victoria (VIC). Black lines in barplots separate samples from different sampling locations. Each coloured bar represents an individual with different colours reflecting estimated proportional membership to particular clusters.

3.3. Population-Level Analysis

At the population level, we detected substantial genetic sub-structuring within each geographic region, with the exception of southern WA (Figure 4d). Three clusters were identified in northern WA separating Kimberley islands, Kimberley mainland and Barrow Island (Figure 4d). We detected multiple admixed clusters in southern WA and the modal K value varied between three and four clusters (Figure S3); in each, however, inland southwest WA was consistently distinct from the Perth population. In South Australia, Kangaroo Island was differentiated from Mount Lofty Ranges, but Fleurieu Peninsula was either assigned to Kangaroo Island (STRUCTURE), admixed between the two (TESS), or assigned to its own cluster (DAPC). In Victoria, the Grampians was differentiated from

Mount Burr in south-east SA, while Lower Glenelg either assigned to the same cluster as Mount Burr (STRUCTURE and TESS) or formed its own cluster (DAPC) (Figure S3).

AMOVA also supported geographic structuring of mtDNA haplotypes when each region was analysed separately and revealed that a large proportion of genetic variation occurred between sampling sites, ranging from 27.5–73.3% ($p < 0.001$ in all cases, Table 1). In South Australia, 73.3% of genetic variation was found among populations, while approximately half of the genetic variation was found among populations in northern WA (48.3%) and Victoria (48.4%). In southern WA, more genetic variation was found within population (72%) indicating greater common shared genetic variation among populations than other regions (Table 1), consistent with patterns identified in cluster analyses.

Table 1. Hierarchical AMOVA of mitochondrial DNA performed by grouping sampling sites into regions. Barrow Island (BWI), Kimberley Island (KIMI), Kimberley mainland (KIMM), south west Western Australia (SW WA), Mount Lofty Ranges (MLR), Fleurieu Peninsula (FP), Kangaroo Island (KI), Lower Glenelg (LG).

Comparison	Source of Variation	Variance Components	Percentage	p-Value
BWI, KIMI, KIMM	Among population	2.1	48.3	<0.001
	Within population	2.3	51.7	
Perth metro, SW WA Coast, SW WA inland	Among population	1.7	27.9	<0.001
	Within population	4.3	72.1	
MLR, FP, KI	Among population	5.6	73.3	<0.001
	Within population	2.0	26.7	
LG, The Grampians, Mount Burr	Among population	1.0	51.6	<0.001
	Within population	1.0	48.4	

Pairwise population D_{est} values based on microsatellite and ϕ_{ST} values based on mtDNA data indicated a substantial genetic divergence between *Isoodon* in different regions (Table 2). The highest differentiation was recorded between SA and Victoria samples with D_{est} and ϕ_{ST} values ranging from 0.87–0.97 and 0.75–0.98, respectively, and comparable to differentiation between Victoria and WA samples (D_{est} 0.69–0.88 and ϕ_{ST} 0.71–0.88). SA and southern WA showed lower genetic differentiation (D_{est} 0.38–0.65 and ϕ_{ST} 0.36–0.61), while *Isoodon* from northern WA and southern WA indicated significant allelic differentiation (D_{est} 0.62–0.82) and comparable differentiation in mtDNA (ϕ_{ST} 0.38–0.68).

Substantial genetic differentiation was also observed amongst populations within geographic regions, particularly of Kangaroo Island from remaining SA populations (Table 2).

3.4. Population Genetic Diversity

Overall, mtDNA diversity was high and comparable between southern WA, northern WA, and SA *Isoodon* populations (Table 3). Across all diversity metrics, Victorian populations had low diversity, with lower haplotype diversity than all other populations (Table 3) and similarly low nuclear diversity to SA ($H = 0.44$–0.61 and 0.42–0.60, respectively; Friedman's test, $p < 0.05$) indicating these populations may be in genetic decline. Multi-locus F_{IS} values had significantly positive F_{IS} values indicating a possible Wahlund effect (samples were sourced from multiple populations) or non-random mating patterns within all sites with the exception of Kimberley mainland (randomization tests, $p < 0.0042$, Table 3). Kimberley mainland had the highest number of unique haplotypes, yet most haplotypes occurred at low frequencies. This pattern is concordant with evidence of recent population expansion according to the unimodal mismatch distribution and correlation with the spatial expansion model (Tajima's D = -1.87, Fu's Fs = -5.90, $p < 0.05$ in both cases; Figure S5). Similarly, the Kimberley mainland retains high microsatellite diversity relative to other populations.

Table 2. Pairwise D_{est} and ϕ_{ST} values between samples from different locations based on microsatellite (above diagonal) and mtDNA data (below diagonal).

State	Sampling Site	WA						SA			VIC		
		Barrow Island	Kimberley Islands	Kimberley Mainland	Perth Metro	SW WA Coast	SW WA Inland	Kangaroo Island	Fleurieu Peninsula	Mount Lofty Ranges	The Grampians	Lower Glenelg	Mount Burr
WA	Barrow Island	–	0.436	0.478	0.701	0.682		0.902	0.817	0.817	0.868	0.862	0.783
	Kimberley islands	0.391	–	0.331	0.791	0.821		0.888	0.895	0.877	0.874	0.876	0.764
	Kimberley mainland	0.495	0.462	–	0.656	0.623		0.788	0.776	0.761	0.785	0.846	0.705
	Perth metro	0.471	0.435	0.579	–	0.111		0.611	0.550	0.423	0.853	0.880	0.792
	SW WA Coast	0.582	0.533	0.685	0.318	–		0.602	0.450	0.359	0.783	0.894	0.778
	SW WA inland	0.437	0.389	0.564	0.262	0.231	–						
SA	Kangaroo Island	0.815	0.711	0.781	0.502	0.646	0.579	–	0.503	0.535	0.894	0.970	0.899
	Fleurieu Peninsula	0.558	0.465	0.66	0.424	0.565	0.375	0.922	–	0.183	0.870	0.971	0.895
	Mount Lofty Ranges	0.400	0.284	0.445	0.415	0.520	0.383	0.754	0.404	–	0.913	0.960	0.896
VIC	Grampians	0.869	0.824	0.843	0.739	0.804	0.756	0.971	0.933	0.828	–	0.303	0.302
	Lower Glenelg	0.877	0.835	0.843	0.751	0.824	0.765	0.981	0.946	0.846	0.765	–	0.177
	Mount Burr	0.778	0.746	0.803	0.703	0.744	0.687	0.899	0.830	0.747	0.555	0.085	–

Significant D_{est} values and ϕ_{ST} values after approximately 1000 permutations are denoted with bold font. Colours represent levels of differentiation from light to dark: 0.0–0.2, 0.2–0.4, 0.4–0.6, 0.6–0.8, and 0.8–1.0.

Table 3. Nuclear and mitochondrial DNA genetic diversity characteristics of *Isoodon* populations.

State	Region	Location	N D-loop	Nh	uh	h	Π	N	H	A_r	A	F_{IS}
WA			159	61	61	0.916 ± 0.017	0.023 ± 0.011	172	0.835 ± 0.030	18.1 ± 2.2	19.0 ± 2.3	
		North WA	87	36	36	0.761 ± 0.051	0.013 ± 0.007	109	0.744 ± 0.064 [A,C]	13.1 ± 1.7 [A]	15.0 ± 2.0 [A]	
		Barrow Island	11	7	7	0.873 ± 0.089	0.010 ± 0.006	77	0.642 ± 0.084	3.2 ± 0.3	8.1 ± 1.3	**0.056**
		Kimberley Island	11	8	8	0.927 ± 0.067	0.015 ± 0.009	16	0.649 ± 0.051	3.1 ± 0.2	7.0 ± 0.9	**0.277**
		Kimberley mainland	65	21	21	0.585 ± 0.075	0.008 ± 0.005	16	0.861 ± 0.030	4.4 ± 0.3	10.8 ± 1.5	0.075
		South WA	71	25	25	0.939 ± 0.014	0.021 ± 0.011	63	0.786 ± 0.044 [A]	12.0 ± 1.7 [A]	12.1 ± 1.8 [A]	
		Perth metro	40	12	10	0.853 ± 0.039	0.018 ± 0.009	39	0.759 ± 0.051	3.8 ± 0.3	9.6 ± 1.0	**0.175**
		South West Coast	13	5	3	0.782 ± 0.079	0.016 ± 0.009	7	0.767 ± 0.071	3.8 ± 0.4	6.0 ± 0.8	**0.231**
		South West inland	19	10	10	0.906 ± 0.041	0.017 ± 0.009	17	0.787 ± 0.030	3.8 ± 0.2	7.6 ± 0.9	**0.128**
SA			21	9	9	0.857 ± 0.056	0.023 ± 0.012	373	0.623 ± 0.050 [B]	5.7 ± 0.7 [B]	6.9 ± 0.8 [B]	
		Mount Lofty Ranges	9	4	4	0.778 ± 0.110	0.015 ± 0.009	311	0.597 ± 0.052	2.7 ± 0.2	5.7 ± 0.8	**0.251**
		Fleurieu Peninsula	4	3	3	0.833 ± 0.222	0.008 ± 0.006	45	0.567 ± 0.056	2.6 ± 0.2	4.4 ± 0.5	**0.458**
		Kangaroo Island	8	2	2	0.250 ± 0.180	0.001 ± 0.001	17	0.422 ± 0.079	2.1 ± 0.3	3.1 ± 0.5	**0.230**
VIC			38	7	7	0.634 ± 0.060	0.008 ± 0.005	186	0.627 ± 0.059 [B,C]	5.8 ± 0.5 [B]	6.7 ± 0.6 [B]	
		Grampians	11	2	1	0.182 ± 0.144	0.002 ± 0.002	27	0.486 ± 0.078	2.3 ± 0.3	3.1 ± 0.4	**0.243**
		Lower Glenelg	12	2	0	0.167 ± 0.134	0.001 ± 0.001	12	0.444 ± 0.098	2.0 ± 0.2	2.4 ± 0.4	**0.353**
		Mount Burr	11	2	1	0.327 ± 0.153	0.009 ± 0.005	147	0.612 ± 0.057	2.9 ± 0.2	6.0 ± 0.6	**0.290**

mtDNA sample size ($N_{\text{D-loop}}$), number of haplotypes (Nh), number of unique haplotypes (uh), haplotype diversity (h), nucleotide diversity (Π), microsatellite sample size (N), gene diversity (H), allelic richness (A_r), allele number (A), and inbreeding coefficients (F_{IS}). Standard errors are given after mean values. F_{IS} estimates significantly greater than zero after correction for multiple comparisons are denoted with bold font. Statistical differences amongst geographic regions for nuclear diversity parameters were tested using a Friedman's test and superscripts identify statistically significant groups as determined by post-hoc multiple comparison of group means. Populations with different lettered superscripts are statistically different a- $p < 0.05$.

4. Discussion

In conservation biology, a taxonomic classification system that closely reflects the biology of a species is essential to guide conservation management actions and increase the chance of success. Incorrect designation of species and subspecies could lead to inappropriate management decisions that can have detrimental effects such as outbreeding depression, maladaptation, or separating populations when genetic augmentation could benefit inbred populations [3,5,67]. Through increased spatial sampling and addition of hyper-variable nuclear markers, our study aimed to provide further resolution to the genetic relationships of *Isoodon* species in Western Australia and South Australia, where current taxonomy does not reflect reported phylogeographic relationships [12]. At the broadest level, increased sampling enabled us to resolve two well-supported mtDNA clades representing WA entities *I. auratus* (northern WA) and *I. fusciventer* (southern WA); however, *I. obesulus* in South Australia remained polyphyletic with SA haplotypes represented in both WA clades, suggesting that incomplete lineage sorting may be confounding phylogenetic analyses across this group. Further highlighting the close relationship amongst these three taxa, analyses of nuclear microsatellite markers indicated a geographic cline stretching from northern WA to South Australia rather than resolving distinct taxonomic entities, with sub-structuring amongst taxa only becoming evident in hierarchical clustering analyses. In all analyses, we detected highly localised patterns of mitochondrial and nuclear genetic diversity indicating limited contemporary gene flow. Taken together, our analyses indicate a complex phylogeographic history for *Isoodon* sp. in southern and western Australia. We discuss identification of conservation units below, as well as some of the underlying processes that may have contributed to the observed results.

4.1. Distinction between I. fusciventer, I. auratus, and I. obesulus (SA)

The relationship of WA taxa *I. auratus* and *I. fusciventer* has long been queried, with multiple mtDNA studies over the past 20 years reporting the taxa as polyphyletic, evidence used to suggest the two species be synonymised [12,15,21]. This observation has been controversial given the striking morphological differences between *I. auratus* and *I. fusciventer*, with *I. fusciventer* more similar in size and pelage colour to *I. obesulus* than *I. auratus* (Figure 2; [19]), although several case-studies have indicated morphological variation in WA bandicoots can be strongly environmentally determined [68–71].

Here, with expanded geographic sampling of both *I. auratus* (northern WA) and *I. fusciventer* (southern WA), we largely resolved two monophyletic groups in mtDNA (putative ESUs) with strong Bayesian support (PP 0.93). However, a very small number of haplotypes from *I. fusciventer* were clustered with *I. auratus* and vice versa, rendering these groups polyphyletic. Further, to add complexity to this relationship, haplotypes from South Australian *I. obesulus* were polyphyletic with *I. auratus* and *I. fusciventer*, with Kangaroo Island and Fleurieu Peninsula samples clustered within *I. fusciventer* and Mount Lofty Ranges samples polyphyletic between *I. auratus* and *I. fusciventer*. Even though SA mtDNA haplotypes were unique, they were most closely related to haplotypes of either *I. auratus* and *I. fusciventer* indicating a relictual connection between these lineages. For species undergoing allopatric speciation, coalescent theory predicts that the process is detected in molecular phylogenies through the progression from a pattern of polyphyly to paraphyly to reciprocal monophyly, the timing of which is a function of effective population size with large populations taking longer to achieve reciprocal monophyly than small ones [72,73]. *Isoodon auratus*, in particular, was historically abundant and widespread across much of central and western Australia, possibly extending into New South Wales [17] before massive range collapse following European settlement, and *I. auratus* and *I. fusciventer* currently retain large population sizes relative to other eastern *I. obesulus* [28]. The close relationships of various SA *I. obesulus*, *I. auratus*, and *I. fusciventer* mtDNA haplotypes likely reflects historical connection amongst these taxa and suggests that time since geographic isolation has been insufficient to allow complete lineage sorting at this locus. This is perhaps unsurprising given the recent diversification of bandicoots in Australia within the late Pleistocene (3.1 Mya), with more recent speciation between *I. auratus* and *I. fusciventer* estimated at approximately 1.5 Mya [19]. An opportunity to genetically sample extinct populations from the previous range of these species (e.g., via museum specimens or sub-fossil deposits [74]) could help to more fully illuminate the historical demography and connectivity across this group.

Analyses of nuclear (microsatellite) allele frequency data showed inconsistency with mtDNA results for the relationship of SA *I. obesulus* with WA taxa. PCoA of microsatellite allele frequencies indicated a geographic cline ranging from northern WA, through southern WA to South Australia reflecting *I. fusciventer* as a connecting population according to their geographic relationship. In contrast, PCoA of mtDNA showed SA *I. obesulus* as closely related to both *I. auratus* and *I. fusciventer* with the latter two species at the ends of the cline. Clustering analyses of microsatellite data (TESS, Structure, DAPC) consistently resolved K = 2 clusters separating WA (*I. auratus, I. fusciventer*) from SA (*I. obesulus*), albeit with a pattern of admixture between the two clusters occurring in southern WA (Structure, TESS) that is often indicative of a pattern of genetic isolation by distance [75,76]. Such a pattern of IBD in microsatellite data may reflect greater male-biased dispersal, whereas restricted female dispersal has led to greater structuring in mitochondrial DNA that is maternally inherited (e.g., [74]). Hierarchical structuring present within the WA cluster was revealed upon stepwise removal of divergent groups in clustering analyses, which then resolved K = 3 clusters within WA representing *I. a. auratus* (Kimberley), *I. a. barrowensis* (Barrow Island), and *I. fusciventer* (southern WA).

Conflicting signals from mitochondrial and nuclear DNA analyses in our study thus make diagnosis of ESUs difficult under Moritz's genetic criteria of reciprocal monophyly, particularly in relation to SA *I. obesulus*. Under the less restrictive Adaptive Evolutionary Conservation concept of Fraser and Bernatchez [8], requiring only demonstration of highly restricted gene flow amongst lineages with a fundamental goal to preserve adaptive genetic variance within species, our data suggest *I. auratus*, *I. fusciventer*, and SA *I. obesulus* represent separate ESUs, with additional hierarchical substructure identified within *I. auratus* separating *I. auratus auratus* (northern WA, Kimberley) and *I. a. barrowensis* (northern WA, Barrow Island). The high genetic distinction of *I. auratus*, *I. fusciventer* and SA *I. obesulus* in mitochondrial and nuclear DNA is reflective of highly reduced gene flow, likely as a result of their allopatric distribution, and is in line with previous results from allozyme data [12] and morphological differences [15,19,22,23]. Multiple lines of evidence suggest these are distinct evolutionary lineages that warrant separate taxonomic status. While SA *I. obesulus* appeared related to both *I. fusciventer* and *I. auratus* in mtDNA, cluster analysis of nuclear data indicated that SA *I. obesulus* is more genetically distinct from the remaining WA taxa, which may warrant recognition of this entity at species level, particularly if *I. auratus* and *I. fusciventer* are retained as such. Allozyme data presented in Cooper et al. [12] shows a much closer relationship of *I. fusciventer* and SA *I. obesulus*, clearly distinct from *I. auratus*, which is consistent at least with the superficial morphology of the two entities. Further genetic and morphological investigation is required to clarify taxon boundaries amongst these entities, particularly as the SA population is highly restricted and declining and likely to require urgent conservation attention.

4.2. Lack of Clarity on the Taxonomic Relationship of SA I. obesulus from Current Genetic Data

Previous studies of the phylogenetic relationships of *Isoodon* bandicoots have relied on small numbers of samples and/or limited numbers of phylogenetic markers (mitochondrial or nuclear sequencing markers) [12,15,20,21]. Here, with expanded sampling of problematic taxa (*I. auratus*, in particular) and with the addition of high mutation rate nuclear markers (microsatellites), we aimed to provide greater resolution to the relationships between purported taxa in the "southern and western" clade of closely related *Isoodon* bandicoots identified in Cooper et al. [12]. Our analyses recovered ESUs and sub-hierarchical ESUs that are largely reflective of current taxonomic designations, although ambiguity remains over the taxonomic identification of SA *I. obesulus*. There is growing recognition that multi-locus and multi-species coalescent approaches are required to overcome the limitations of current species delimitation methods, particularly when single gene trees are poorly resolved or in conflict with the actual species tree, as may arise due to ancestral lineage sorting, hybridisation, and a range of other population genetic processes [77,78]. Here, we find a disconnect in the results obtained for mitochondrial and nuclear DNA analyses for SA *I. obesulus* that may be attributed to a mismatch in the tempo of mutation rates in the markers we have used to the evolutionary processes occurring in the group. For example, mtDNA analysis is impacted by incomplete lineage sorting, and microsatellites may have mutation rates that are too rapid to reflect speciation patterns, as they largely reflect contemporary population genetic processes. Genomic sequencing approaches, for example, targeted exon capture [79] enable concurrent sampling of multiple, genome-wide markers with a range of mutation rates and could provide sufficient resolution to resolve speciation patterns across southern and western Australia. Given the relatively recent pattern of speciation and massive range collapse experienced by these taxa investigation of the demographic history of the group will be particularly fascinating, especially as the current distribution boundaries of *Isoodon* populations are coincident with biogeographic barriers that have been identified elsewhere (e.g., Nullarbor barrier, Murravian barrier; [80]).

4.3. Genetic Diversity and Population Substructure

Bandicoot species, particularly across southern Australia, are highly impacted by habitat fragmentation, habitat loss, and declining population size. Here, we found that both Victorian and South Australian populations of *I. obesulus* showed consistently low genetic diversity across nuclear markers relative to WA populations, and that Victorian populations had low mitochondrial diversity. In contrast, the Kimberley mainland population of *I. auratus,* which is less impacted by anthropogenic fragmentation effects, retains high genetic diversity with signals suggesting population expansion in the region.

Consistent with this, we also found evidence of restricted gene flow with very high numbers of unique local mtDNA haplotypes, significant and high pairwise D_{est} and ϕ_{ST} values, and a significant proportion of genetic variation apportioned between, rather than within, populations. We also detected significant genetic sub-structuring within each region via cluster analyses, although populations in southern WA were more admixed than elsewhere. Such a pattern is consistent with the expected limited dispersal range of bandicoots (~3 km; Li et al. [81,82]), but connectivity is also likely moderated by the amount (and fragmentation) of native vegetation in the landscape [13].

4.4. Recommendations for Conservation Management

Our mtDNA analysis, haplotype uniqueness, and significant divergence among populations in different regions suggested that *I. auratus, I. fusciventer*, and *I. obesulus* in South Australia should be managed as separate units, and that SA *I. obesulus* requires urgent consideration to assess its taxonomy and conservation status. Several populations included in this study showed evidence of genetic decline, particularly those in Victoria and South Australia. Ongoing habitat fragmentation and degradation in these regions may be contributing to limited gene flow amongst patches and subsequently genetic drift. Improving landscape-scale connectivity will be of high priority for these populations with evidence that low, shrubby vegetation can provide a structural habitat for bandicoots to avoid predation as well as connectivity, even in degraded environments [13,14,83]. Management of introduced predators (foxes, cats) and habitat quality improvement will also be critical factors [84–86]. As one of Australia's important digging mammals, bandicoots are also targeted for translocation to restore ecological function in ecological restoration projects and to secure "safe haven" sites to improve their conservation status [87–93]. Golden bandicoots (*I. auratus*) are of particular importance for arid zone translocations with those undertaken to date considered successful [30,70]. There is evidence that bandicoots show morphological adaptation to local conditions [68–70], which has been shown to have a genetic, as well as an environmental component [71], suggesting that translocation to similar habitats is important. However, all translocations to date have been in north-western Australia and have involved the Barrow Island population of *I. auratus*, which we demonstrate here to have lower genetic diversity (being an island population) than mainland Kimberley *I. auratus*. It may be prudent when planning future translocations to consider whether it is feasible to source animals from Kimberley populations to ensure genetic diversity within the species is maintained across multiple sites to reduce risks to further genetic decline.

Supplementary Materials: The following are available online at https://www.mdpi.com/1424-2818/13/1/2/s1.

Author Contributions: Conceptualization, R.T., K.O., S.J.B.C., and M.B.; methodology, R.T. and K.O.; software, R.T. and K.O.; validation, R.T. and K.O.; formal analysis, R.T.; investigation, R.T. and E.L.; resources, S.J.B.C., Y.L., and K.O.; data curation, R.T.; writing—original draft preparation, R.T. and K.O.; writing—review and editing, K.O., S.J.B.C., and M.B.; visualization, R.T. and K.O.; supervision, K.O. and M.B.; project administration, R.T.; funding acquisition, K.O. All authors have read and agreed to the published version of the manuscript.

Funding: No specific funding was provided for this research.

Acknowledgments: We thank A. Wayne, P. Glen, G. Harewood, P. Townsing for provision of additional *I. fusciventer* samples for this analysis and J. Smith and staff from the Australian Wildlife Conservancy for additional *I. auratus* samples from the Kimberley. Contributors to original datasets obtained from [12,13,20,30] are indicated in Acknowledgements of the respective publications.

Conflicts of Interest: The authors declare no conflict of interest.

References

1. Frankham, R. Challenges and opportunities of genetic approaches to biological conservation. *Biol. Conserv.* **2010**, *143*, 1919–1927. [CrossRef]
2. Moritz, C. Strategies to protect biological diversity and the evolutionary processes that sustain it. *Syst. Biol.* **2002**, *51*, 238–254. [CrossRef] [PubMed]
3. Frankham, R.; Ballou, J.D.; Eldridge, M.D.B.; Lacy, R.C.; Ralls, K.; Dudash, M.R.; Fenster, C.B. Predicting the probability of outbreeding depression. *Conserv. Biol.* **2011**, *25*, 465–475. [CrossRef] [PubMed]
4. Allendorf, F.W.; Leary, R.F.; Spruell, P.; Wenburg, J.K. The problems with hybrids: Setting conservation guidelines. *Trends Ecol. Evol.* **2001**, *16*, 613–622. [CrossRef]
5. Weeks, A.R.; Stoklosa, J.; Hoffmann, A.A. Conservation of genetic uniqueness of populations may increase extinction likelihood of endangered species: The case of Australian mammals. *Front. Zoöl.* **2016**, *13*, 1–9. [CrossRef]
6. Ryder, O.A. Species conservation and systematics: The dilemma of subspecies. *Trends Ecol. Evol.* **1986**, *1*, 9–10. [CrossRef]
7. Moritz, C. Defining 'evolutionarily significant units' for conservation. *Trends Ecol. Evol.* **1994**, *9*, 373–375. [CrossRef]
8. Fraser, D.J.; Bernatchez, L. Adaptive evolutionary conservation: Towards a unified concept for defining conservation units. *Mol. Ecol.* **2001**, *10*, 2741–2752. [CrossRef]
9. Crandall, K.A.; Bininda-Emonds, O.R.; Mace, G.M.; Wayne, R.K. Considering evolutionary processes in conservation biology. *Trends Ecol. Evol.* **2000**, *15*, 290–295. [CrossRef]
10. Gordon, G.; Hulbert, A.J. Peramelidae. In *Fauna of Australia. Mammalia*; Walton, D.W., Richardson, B.J., Eds.; Australian Government Publishing Service: Canberra, Australia, 1989; Volume 1b, pp. 603–624.
11. Warburton, N.M.; Travouillon, K.J. The biology and palaeontology of the Peramelemorphia: A review of current knowledge and future research directions. *Aust. J. Zoöl.* **2016**, *64*, 151–181. [CrossRef]
12. Cooper, S.; Ottewell, K.; Macdonald, A.J.; Adams, M.; Byrne, M.; Carthew, S.M.; Eldridge, M.D.B.; Li, Y.; Pope, L.C.; Saint, K.M.; et al. Phylogeography of southern brown and golden bandicoots: Implications for the taxonomy and distribution of endangered subspecies and species. *Aust. J. Zoöl.* **2018**, *66*, 379. [CrossRef]
13. Ottewell, K.; Pitt, G.; Pellegrino, B.; Van Dongen, R.; Kinloch, J.; Willers, N.; Byrne, M. Remnant vegetation provides genetic connectivity for a critical weight range mammal in a rapidly urbanising landscape. *Landsc. Urban Plan.* **2019**, *190*, 103587. [CrossRef]
14. Maclagan, S.J.; Coates, T.; Ritchie, E.G. Don't judge habitat on its novelty: Assessing the value of novel habitats for an endangered mammal in a peri-urban landscape. *Biol. Conserv.* **2018**, *223*, 11–18. [CrossRef]
15. Pope, L.C.; Storch, D.; Adams, M.; Moritz, C.; Gordon, G. A phylogeny for the genus *Isoodon* and a range extension for *I. obesulus peninsulae* based on mtDNA control region and morphology. *Aust. J. Zoöl.* **2001**, *49*, 411. [CrossRef]
16. Wilson, D.E.; Reeder, D.M. *Mammal Species of the World*, 3rd ed.; A Taxonomic and Geographic Reference; Johns Hopkins University Press: Baltimore, MD, USA, 2005.

17. Van Dyck, S.; Strahan, R. *The Mammals of Australia*, 3rd ed.; Reed New Holland: Sydney, Australia, 2008.

18. Westerman, M.; Kear, B.; Aplin, K.; Meredith, R.; Emerling, C.; Springer, M. Phylogenetic relationships of living and recently extinct bandicoots based on nuclear and mitochondrial DNA sequences. *Mol. Phylogenetics Evol.* **2012**, *62*, 97–108. [CrossRef]

19. Travouillon, K.J.; Phillips, M.J. Total evidence analysis of the phylogenetic relationships of bandicoots and bilbies (Marsupialia: Peramelemorphia): Reassessment of two species and description of a new species. *Zootaxa* **2018**, *4378*, 224–256. [CrossRef]

20. Li, Y.; Lancaster, M.L.; Carthew, S.M.; Packer, J.G.; Cooper, S.J.B. Delineation of conservation units in an endangered marsupial, the southern brown bandicoot (*Isoodon obesulus obesulus*), in South Australia/western Victoria, Australia. *Aust. J. Zoöl.* **2014**, *62*, 345–359. [CrossRef]

21. Zenger, K.R.; Eldridge, M.D.B.; Johnston, P.G. Phylogenetics, population structure and genetic diversity of the endangered southern brown bandicoot (*Isoodon obesulus*) in south-eastern Australia. *Conserv. Genet.* **2005**, *6*, 193–204. [CrossRef]

22. Lyne, A.G.; Mort, P.A. A comparison of skull morphology in the marsupial bandicoot genus *Isoodon*: Its taxonomic implications and notes on a new species, *Isoodon arnhemensis*. *Aust. Mammal.* **1981**, *4*, 107–133.

23. Menkhorst, P.; Knight, F. *A Field Guide to the Mammals of Australia*; Oxford University Press: Melbourne, Australia, 2011.

24. Coates, T.; Nicholls, D.; Willig, R. The distribution of the southern brown bandicoot 'Isoodon obesulus' in south central Victoria. *Vic. Nat.* **2008**, *125*, 128–139.

25. Paull, D.J.; Mills, D.J.; Claridge, A.W. Fragmentation of the southern brown bandicoot *Isoodon obesulus*: Unraveling past climate change from vegetation clearing. *Int. J. Ecol.* **2013**, *2013*, 536524. [CrossRef]

26. Burbidge, A.A.; Woinarski, J. *Isoodon auratus*. The IUCN Red List of Threatened Species 2016; Vol. e.T10863A115100163. Available online: http://dx.doi.org/10.2305/IUCN.UK.2016-3.RLTS.T10863A21966258.en (accessed on 23 July 2019).

27. Burbidge, A.A.; Woinarski, J. *Isoodon obesulus*. The IUCN Red List of Threatened Species 2016; Vol. e.T40553A115173603. Available online: http://dx.doi.org/10.2305/IUCN.UK.2016-3.RLTS.T40553A21966368.en (accessed on 26 March 2019).

28. Woinarski, J.C.Z.; Burbidge, A.A.; Harrison, P.; Kelly, J. *The Action Plan for Australian Mammals 2012*; CSIRO Publishing: Collingwood, Australia, 2014.

29. Department of the Environment. Creating Safe Havens. Commonwealth of Australia. 2015. Available online: https://www.environment.gov.au/system/files/resources/f3e6ed38-6b27-46ca-a980-66ce3a1cdadc/files/factsheet-creating-safe-havens.pdf (accessed on 27 March 2019).

30. Ottewell, K.; Dunlop, J.; Thomas, N.; Morris, K.; Coates, D.; Byrne, M. Evaluating success of translocations in maintaining genetic diversity in a threatened mammal. *Biol. Conserv.* **2014**, *171*, 209–219. [CrossRef]

31. Fumagalli, L.; Pope, L.C.; Taberlet, P.; Moritz, C. Versatile primers for the amplification of the mitochondrial DNA control region in marsupials. *Mol. Ecol.* **1997**, *6*, 1199–1201. [CrossRef] [PubMed]

32. Thompson, J.D.; Gibson, T.J.; Plewniak, F.; Jeanmougin, F.; Higgins, D.G. The CLUSTAL_X windows interface: Flexible strategies for multiple sequence alignment aided by quality analysis tools. *Nucleic Acids Res.* **1997**, *25*, 4876–4882. [CrossRef]

33. Kumar, S.; Stecher, G.; Li, M.; Knyaz, C.; Tamura, K. MEGA X: Molecular Evolutionary Genetics Analysis across Computing Platforms. *Mol. Biol. Evol.* **2018**, *35*, 1547–1549. [CrossRef]

34. Stamatakis, A. RAxML-VI-HPC: Maximum likelihood-based phylogenetic analyses with thousands of taxa and mixed models. *Bioinformatics* **2006**, *22*, 2688–2690. [CrossRef]

35. Suchard, M.A.; Lemey, P.; Baele, G.; Ayres, D.L.; Drummond, A.J.; Rambaut, A. Bayesian phylogenetic and phylodynamic data integration using BEAST 1.10. *Virus Evol.* **2018**, *4*, vey016. [CrossRef]

36. Rodríguez, F.; Oliver, J.; Marín, A.; Medina, J. The general stochastic model of nucleotide substitution. *J. Theor. Biol.* **1990**, *142*, 485–501. [CrossRef]

37. Yang, Z. Among-site rate variation and its impact on phylogenetic analyses. *Trends Ecol. Evol.* **1996**, *11*, 367–372. [CrossRef]

38. Darriba, D.; Taboada, G.L.; Doallo, R.; Posada, D. jModelTest 2: More models, new heuristics and parallel computing. *Nat. Methods* **2012**, *9*, 772. [CrossRef]

39. Rambaut, A.; Drummond, A.J.; Xie, D.; Baele, G.; Suchard, M.A. Posterior summarization in bayesian phylogenetics using Tracer 1.7. *Syst. Biol.* **2018**, *67*, 901–904. [CrossRef] [PubMed]

40. Paradis, E.; Schliep, K. ape 5.0: An environment for modern phylogenetics and evolutionary analyses in R. *Bioinformatics* **2018**, *35*, 526–528. [CrossRef] [PubMed]

41. Revell, L.J. phytools: An R package for phylogenetic comparative biology (and other things). *Methods Ecol. Evol.* **2011**, *3*, 217–223. [CrossRef]

42. R Core Team. *R: A Language and Environment for Statistical Computing*; R Foundation for Statistical Computing: Vienna, Austria, 2019.

43. Dray, S.; Dufour, A.-B. The ade4 package: Implementing the duality diagram for ecologists. *J. Stat. Softw.* **2007**, *22*, 1–20. [CrossRef]

44. Jombart, T. Adegenet: A R package for the multivariate analysis of genetic markers. *Bioinformatics* **2008**, *24*, 1403–1405. [CrossRef]

45. Pritchard, J.K.; Stephens, M.; Donnelly, P. Inference of population structure using multilocus genotype data. *Genetics* **2000**, *155*, 945–959.

46. Evanno, G.; Regnaut, S.; Goudet, J. Detecting the number of clusters of individuals using the software structure: A simulation study. *Mol. Ecol.* **2005**, *14*, 2611–2620. [CrossRef]

47. Earl, D.A.; Vonholdt, B.M. STRUCTURE HARVESTER: A website and program for visualizing STRUCTURE output and implementing the Evanno method. *Conserv. Genet. Resour.* **2012**, *4*, 359–361. [CrossRef]

48. Jakobsson, M.; Rosenberg, N.A. CLUMPP: A cluster matching and permutation program for dealing with label switching and multimodality in analysis of population structure. *Bioinformatics* **2007**, *23*, 1801–1806. [CrossRef]

49. Chen, C.; Durand, E.; Forbes, F.; François, O. Bayesian clustering algorithms ascertaining spatial population structure: A new computer program and a comparison study. *Mol. Ecol. Notes* **2007**, *7*, 747–756. [CrossRef]

50. François, O.; Ancelet, S.; Guillot, G. Bayesian clustering using hidden markov random fields in spatial population genetics. *Genetics* **2006**, *174*, 805–816. [CrossRef]

51. Jombart, T.; Devillard, S.; Balloux, F. Discriminant analysis of principal components: A new method for the analysis of genetically structured populations. *BMC Genet.* **2010**, *11*, 94. [CrossRef] [PubMed]

52. Excoffier, L.; Laval, G.; Schneider, S. Arlequin (version 3.0): An integrated software package for population genetics data analysis. *Evol. Bioinform.* **2005**, *1*, 47–50. [CrossRef]

53. Jost, L. Gst and its relatives do not measure differentiation. *Mol. Ecol.* **2008**, *17*, 4015–4026. [CrossRef] [PubMed]

54. Peakall, R.; Smouse, P.E. GenAlEx 6.5: Genetic analysis in Excel. Population genetic software for teaching and research—An update. *Bioinformatics* **2012**, *28*, 2537–2539. [CrossRef] [PubMed]

55. Fu, Y.X. Statistical tests of neutrality of mutations against population growth, hitchhiking and background selection. *Genetics* **1997**, *147*, 915–925. [PubMed]

56. Tajima, F. Statistical method for testing the neutral mutation hypothesis by DNA polymorphism. *Genetics* **1989**, *123*, 585–595.

57. Slatkin, M.; Hudson, R.R. Pairwise comparisons of mitochondrial DNA sequences in stable and exponentially growing populations. *Genetics* **1991**, *129*, 555–562.

58. Rogers, A.R.; Harpending, H. Population growth makes waves in the distribution of pairwise genetic differences. *Mol. Biol. Evol.* **1992**, *9*, 552–569. [CrossRef]

59. Ray, N.; Currat, M.; Excoffier, L. Intra-deme molecular diversity in spatially expanding populations. *Mol. Biol. Evol.* **2003**, *20*, 76–86. [CrossRef]

60. Excoffier, L. Patterns of DNA sequence diversity and genetic structure after a range expansion: Lessons from the infinite-island model. *Mol. Ecol.* **2004**, *13*, 853–864. [CrossRef] [PubMed]

61. Pompanon, F.; Bonin, A.; Bellemain, E.; Taberlet, P. Genotyping errors: Causes, consequences and solutions. *Nat. Rev. Genet.* **2005**, *6*, 847–859. [CrossRef] [PubMed]

62. Van Oosterhout, C.; Hutchinson, W.F.; Wills, D.P.M.; Shipley, P. Micro-checker: Software for identifying and correcting genotyping errors in microsatellite data. *Mol. Ecol. Notes* **2004**, *4*, 535–538. [CrossRef]

63. Goudet, J. Fstat, a Program to Estimate and Test Gene Diversities and Fixation Indices Version 2.9.3. 2001. Available online: http://www2.unil.ch/popgen/softwares/fstat.htm (accessed on 1 November 2018).

64. Pohlert, T. *The Pairwise Multiple Comparison of Mean Ranks Package (PMCMR)*; R package: Madison, WI, USA, 2014.

65. Australia Wildlife Conservancy. No Data. Available online: http://www.australianwildlife.org/media/59016/golden-bandicoot-artesian-range-sanctuary-photo-ross-knowles-dsc_7435.jpg (accessed on 16 October 2020).

66. Museums Victoria. Isoodon Obesulus Southern Brown Bandicoot in Museums Victoria Collection. 2017. Available online: https://collections.museumsvictoria.com.au/species/8401 (accessed on 9 December 2020).

67. Frankham, R. Genetic rescue of small inbred populations: Meta-analysis reveals large and consistent benefits of gene flow. *Mol. Ecol.* **2015**, *24*, 2610–2618. [CrossRef]

68. Cooper, M.L. Geographic variation in size and shape in the southern brown bandicoot, *Isoodon obesulus* (Peramelidae: Marsupialia), in Western Australia. *Aust. J. Zool.* **1998**, *46*, 145–152. [CrossRef]

69. Cooper, M.L. Temporal variation in skull size and shape in the southern brown bandicoot, *Isoodon obesulus* (Peramelidae: Marsupialia) in Western Australia. *Aust. J. Zool.* **2000**, *48*, 47. [CrossRef]

70. Dunlop, J.; Morris, K. Environmental determination of body size in mammals: Rethinking 'island dwarfism' in the golden bandicoot. *Austral Ecol.* **2018**, *43*, 817–827. [CrossRef]

71. Hale, M. Inheritance of geographic variation in body size, and countergradient variation in growth rates, in the southern brown bandicoot *Isoodon obesulus*. *Aust. Mammal.* **2000**, *22*, 9–16. [CrossRef]

72. Avise, J.C. Phylogeography: Retrospect and prospect. *J. Biogeogr.* **2009**, *36*, 3–15. [CrossRef]

73. Kingman, J.F. Origins of the coalescent. 1974–1982. *Genetics* **2000**, *156*, 1461–1463.

74. Pacioni, C.; Hunt, H.; Allentoft, M.E.; Vaughan, T.G.; Wayne, A.F.; Baynes, A.; Haouchar, D.; Dortch, J.; Bunce, M. Genetic diversity loss in a biodiversity hotspot: Ancient DNA quantifies genetic decline and former connectivity in a critically endangered marsupial. *Mol. Ecol.* **2015**, *24*, 5813–5828. [CrossRef] [PubMed]

75. Czarnomska, S.D.; Niedziałkowska, M.; Borowik, T.; Jędrzejewska, B. Regional and local patterns of genetic variation and structure in yellow-necked mice—The roles of geographic distance, population abundance, and winter severity. *Ecol. Evol.* **2018**, *8*, 8171–8186. [CrossRef] [PubMed]

76. Grosser, S.; Abdelkrim, J.; Wing, J.; Robertson, B.C.; Gemmell, N.J. Strong isolation by distance argues for separate population management of endangered blue duck (*Hymenolaimus malacorhynchos*). *Conserv. Genet.* **2016**, *18*, 327–341. [CrossRef]

77. Yang, Z.; Rannala, B. Bayesian species delimitation using multilocus sequence data. *Proc. Natl. Acad. Sci. USA* **2010**, *107*, 9264–9269. [CrossRef]

78. Knowles, L.L.; Carstens, B.C. Delimiting species without monophyletic gene trees. *Syst. Biol.* **2007**, *56*, 887–895. [CrossRef]

79. Bragg, J.G.; Potter, S.; Bi, K.; Catullo, R.A.; Donnellan, S.; Eldridge, M.D.B.; Joseph, L.; Keogh, J.S.; Oliver, P.M.; Rowe, K.C.; et al. Resources for phylogenomic analyses of Australian terrestrial vertebrates. *Mol. Ecol. Resour.* **2017**, *17*, 869–876. [CrossRef]
80. Dolman, G.; Joseph, L. Evolutionary history of birds across southern Australia: Structure, history and taxonomic implications of mitochondrial DNA diversity in an ecologically diverse suite of species. *Emu Austral Ornithol.* **2015**, *115*, 35–48. [CrossRef]
81. Li, Y.; Cooper, S.J.B.; Lancaster, M.L.; Packer, J.G.; Carthew, S.M. Comparative population genetic structure of the endangered southern brown bandicoot, *Isoodon obesulus*, in fragmented landscapes of Southern Australia. *PLoS ONE* **2016**, *11*, e0152850. [CrossRef]
82. Li, Y.; Lancaster, M.L.; Cooper, S.J.B.; Taylor, A.C.; Carthew, S.M. Population structure and gene flow in the endangered southern brown bandicoot (*Isoodon obesulus obesulus*) across a fragmented landscape. *Conserv. Genet.* **2014**, *16*, 331–345. [CrossRef]
83. Packer, J.G.; Delean, S.; Kueffer, C.; Prider, J.; Abley, K.; Facelli, J.M.; Carthew, S.M. Native faunal communities depend on habitat from non-native plants in novel but not in natural ecosystems. *Biodivers. Conserv.* **2016**, *25*, 503–523. [CrossRef]
84. Lohr, C.; Algar, D. Managing feral cats through an adaptive framework in an arid landscape. *Sci. Total. Environ.* **2020**, *720*, 137631. [CrossRef] [PubMed]
85. Moseby, K.E.; Read, J.; Paton, D.C.; Copley, P.; Hill, B.; Crisp, H. Predation determines the outcome of 10 reintroduction attempts in arid South Australia. *Biol. Conserv.* **2011**, *144*, 2863–2872. [CrossRef]
86. Gardiner, R.; Bain, G.; Hamer, R.; Jones, M.E.; Johnson, C.N. Habitat amount and quality, not patch size, determine persistence of a woodland-dependent mammal in an agricultural landscape. *Landsc. Ecol.* **2018**, *33*, 1837–1849. [CrossRef]
87. Palmer, C. *All Good—Wankura have 2 New Island Homes*; NT Department of Natural Resources, Environment, the Arts and Sport: Darwin, Australia, 2009.
88. Department of Land Resource Management. Nt Action Plan. Golden Bandicoot *Isoodon auratus*. 2015. Available online: https://nt.gov.au/__data/assets/pdf_file/0011/269147/nt-action-plan-golden-bandicoot.pdf (accessed on 26 March 2019).
89. Brown, G.W.; Main, M.L. *National Recovery Plan for the Southern Brown Bandicoot Isoodon obesulus obesulus*; Victorian Government Department of Sustainability and Environment (DSE): Melbourne, Australia, 2010.
90. Haby, N.; Long, K. *Recovery Plan for the Southern Brown Bandicoot in the Mount Lofty Ranges, South Australia, 2004–2009*; Department of Environment and Heritage: Adelaide, South Australia, Australia, 2005.
91. Department of Environment and Conservation. *Gorgon Gas Development—Threatened and Priority Species Translocation and Reintroduction Program—Annual Report 2010/11*; Department of Environment and Conservation: Perth, Australia, 2011.
92. Department of Environment and Conservation. *Barrow Group Nature Reserves Draft Management Plan 2011*; Department of Environment and Conservation: Perth, Australia, 2011.
93. Department of Environment and Conservation (NSW). *Southern Brown Bandicoot* (Isoodon obesulus) *Recovery Plan*; NSW DEC: Hurstville, New South Wales, Australia, 2006.

 *diversity*

Article

Genetic Consequences of Multiple Translocations of the Banded Hare-Wallaby in Western Australia

Daniel J. White [1,*], Kym Ottewell [1,2,3], Peter B. S. Spencer [3], Michael Smith [4,5], Jeff Short [3,6], Colleen Sims [2] and Nicola J. Mitchell [1]

[1] School of Biological Sciences, The University of Western Australia, 35 Stirling Hwy, Crawley, Perth, WA 6009, Australia; kym.ottewell@dbca.wa.gov.au (K.O.); nicola.mitchell@uwa.edu.au (N.J.M.)

[2] Biodiversity and Conservation Science, Department of Biodiversity, Conservation and Attractions, 17 Dick Perry Avenue, Kensington, Perth, WA 6151, Australia; colleen.sims@dbca.wa.gov.au

[3] Environmental and Conservation Sciences, Murdoch University, 90 South Street, Murdoch, Perth, WA 6150, Australia; P.Spencer@murdoch.edu.au (P.B.S.S.); jeff@wildliferesearchmanagement.com.au (J.S.)

[4] The Australian Wildlife Conservancy, P.O. Box 8070 Subiaco East, Perth, WA 6008, Australia; Michael.Smith@australianwildlife.org

[5] Ecosystem Restoration and Intervention Ecology Research Group, School of Biological Sciences, The University of Western Australia, 35 Stirling Highway, Crawley, Perth, WA 6009, Australia

[6] Wildlife Research and Management Pty Ltd., P.O. Box 1360, Kalamunda, WA 6926, Australia

* Correspondence: daniel.white@uwa.edu.au; Tel.: +61-8-6488-1967

Received: 9 October 2020; Accepted: 24 November 2020; Published: 27 November 2020

Abstract: Many Australian mammal species now only occur on islands and fenced mainland havens free from invasive predators. The range of one species, the banded hare-wallaby (*Lagostrophus fasciatus*), had contracted to two offshore islands in Western Australia. To improve survival, four conservation translocations have been attempted with mixed success, and all occurred in the absence of genetic information. Here, we genotyped seven polymorphic microsatellite markers in two source (Bernier Island and Dorre Island), two historic captive, and two translocated *L. fasciatus* populations to determine the impact of multiple translocations on genetic diversity. Subsequently, we used population viability analysis (PVA) and gene retention modelling to determine scenarios that will maximise demographic resilience and genetic richness of two new populations that are currently being established. One translocated population (Wadderin) has undergone a genetic bottleneck and lost 8.1% of its source population's allelic diversity, while the other (Faure Island) may be inbred. We show that founder number is a key parameter when establishing new *L. fasciatus* populations and 100 founders should lead to high survival probabilities. Our modelling predicts that during periodic droughts, the recovery of source populations will be slower post-harvest, while 75% more animals—about 60 individuals—are required to retain adequate allelic diversity in the translocated population. Our approach demonstrates how genetic data coupled with simulations of stochastic environmental events can address central questions in translocation programmes.

Keywords: genetic diversity; population viability analysis; allele retention; translocation; conservation management; threatened marsupial; remnant

---

1. Introduction

Translocation, the anthropogenic movement of a group of organisms from one location to another, is an increasingly necessary tool for conservation management [1]. Conservation translocations take many forms (reintroductions, reinforcements, genetic rescue, assisted colonisation) and in all cases should be carefully planned, as they are costly and their success is hard to predict [2–7]. For example, in fauna translocations, animals may be naïve to predators at translocation sites [6,8], and the harvesting of source populations may disrupt existing social dynamics [9–11]. Nevertheless, translocations are often the only option outside of ex situ conservation, to spread demographic risk and prevent extinctions [12–14].

As translocated populations often arise from a small number of founders, they risk losing genetic diversity via bottlenecks and/or genetic drift, and inbreeding depression can occur due to mating between related individuals [15]. It is therefore important to quantify the genetic diversity in source populations in order to manage these risks and to increase the evolutionary potential of translocated populations [7,16–19]. Individual-based population viability analysis modelling is a powerful tool for making predictions about potential outcomes of various translocation scenarios and can be used to optimise inherent, and often sensitive, trade-offs [18,20]. The incorporation of genetic data into population viability modelling is an important component of conservation decision making, and there has been a marked uptake of these types of analyses for Australian mammals [4,20–23].

Australia has the world's highest rate of mammal extinction [24]. Since European colonisation in 1788, 29 species have gone extinct and the western long-beaked echidna (*Zaglossus bruijnii*) is now extinct in Australia, equating to a rate of loss of around 0.13 species per year [24]. Major threats include predation from exotic introduced pests, in particular the European red fox (*Vulpes vulpes*) and feral cat (*Felis catus*), as well as habitat degradation. Habitat degradation includes both the loss of habitat and habitat fragmentation, both of which can exacerbate predation issues. In addition, small- to medium-sized mammals within a critical weight range (35–5500 g) are more prone to extinction than those outside this range [25–28]. For many of the endemic mammals of Australia that remain, their distributions have contracted to isolated populations, including offshore islands that remain feral predator free. To compound matters, climate change increases the risks of starvation, drought stress, and hyperthermia [29,30]. Accordingly, it is widely agreed that interventions are essential to secure Australia's remaining mammal diversity [24].

The banded hare-wallaby, *Lagostrophus fasciatus*, is a medium-sized (approx. 1700 g), critical weight range herbivorous and nocturnal macropod, the sole member of the Lagostrophinae sub-family. It is currently listed as Vulnerable by the IUCN and under Australia's environmental legislation (the Environment Protection and Biodiversity Conservation Act). Their pre-European range stretched from the coast of central Western Australia to southern South Australia, but the species now survives only in the Shark Bay region of Western Australia on Bernier Island (4267 ha) and Dorre Island (5163 ha) (Figure 1a). The last recorded sighting on the mainland was in 1906 [31]. A 2016 survey suggested there were 2790 individuals on Bernier Island and 2440 individuals on Dorre Island [32], however, populations cycle through boom and bust phases, triggered by rainfall and drought, respectively, and may be reduced by as much as 75% before subsequent recovery [33,34]. These small island populations are expected to have low genetic diversity, which could be further reduced if used as source populations for conservation translocations.

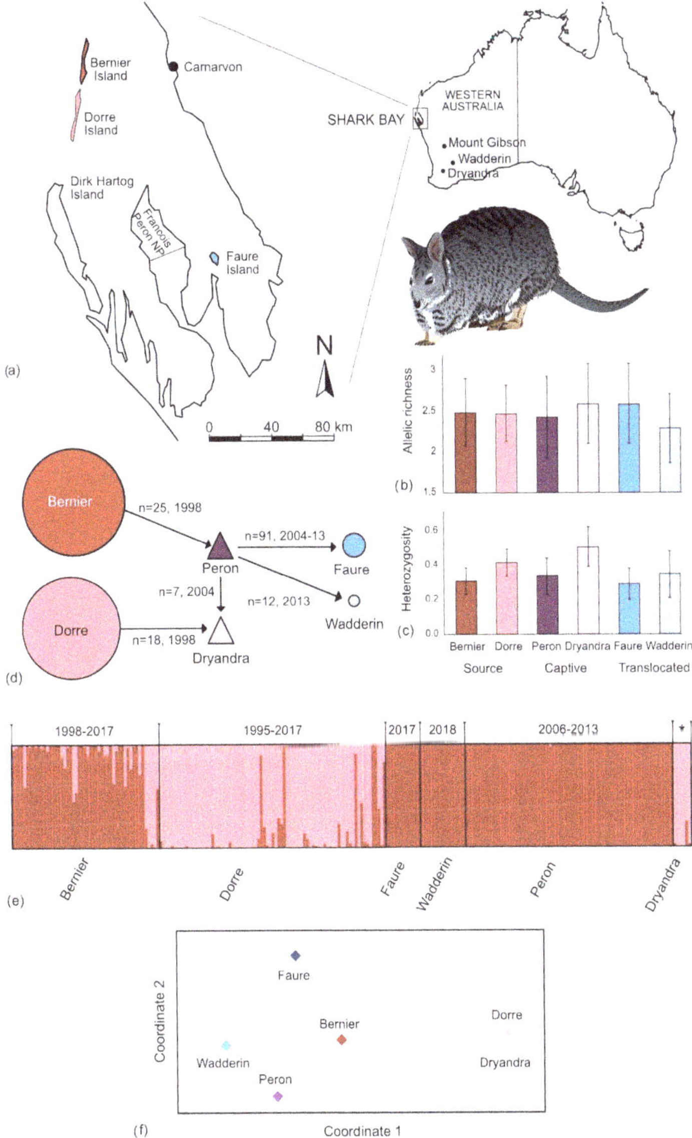

Figure 1. (**a**) Distribution of *L. fasciatus* in Western Australia (inset shows the approximate historical distribution); (**b**,**c**) measures of genetic diversity based on seven microsatellite markers; (**d**) translocation history (failed translocations not shown). Two recent translocations to Mt. Gibson and Dirk Hartog Island are underway; (**e**) STRUCTURE analysis of the two remnant wild populations (Bernier and Dorre Islands), two translocated populations (Faure Island and Wadderin), and two historic captive breeding populations (Peron Captive Breeding Centre (CBC) and Dryandra). Sampling periods are indicated at top of plot; * sampling period for Dryandra is 1999–2002; (**f**) population-level principal coordinates analysis based on genetic distance. The first two axes explain 91.6 of the variation. The *L. fasciatus* image was designed by Creazilla (creazilla.com).

Since 1974, four translocations of *L. fasciatus* have been attempted, and two captive breeding colonies have been established but since discontinued (Figure 1d). The captive colony at Peron Captive Breeding Centre (hereafter Peron CBC) was managed by the Western Australian Department of Biodiversity, Conservation and Attractions (DBCA), and was sourced from Bernier Island. In contrast, a captive breeding colony at Dryandra was sourced from Dorre Island. Two translocation attempts were unsuccessful—specifically the movement of 21 animals from Dorre Island to southern Dirk Hartog Island from 1974 to 1978, and a movement of 18 animals from the Peron CBC to Francois Peron National Park in 2001. The failure of these translocations was attributed to predation by feral cats, drought (Dirk Hartog Island translocation), and the impact of livestock [35–37]. The two successful translocations were each founded from the Peron CBC, with 91 individuals released onto Faure Island between 2004 and 2013, and 12 individuals into Wadderin Sanctuary in 2013. Faure Island is a 4561-hectare island in Shark Bay managed by the Australian Wildlife Conservancy (AWC) where all feral predators and livestock have been removed and was reported to support a population size of approximately 300 in 2017 [38]. Wadderin Sanctuary on the mainland (Figure 1d) is a 430-hectare predator-free enclosure managed by a community group [39] and now holds approximately 30 animals [40].

Currently, two new populations of *L. fasciatus* are being established. From 2017 to 2018, a population of *L. fasciatus* was reintroduced to a 7832 ha safe haven (a fenced area from which introduced predators have been removed) at Mount Gibson Wildlife Sanctuary. A total of 119 individuals were translocated to Mt. Gibson from Bernier ($n = 40$), Faure ($n = 19$) and Dorre ($n = 60$) Islands. In 2017, a translocation programme began for *L. fasciatus* to Dirk Hartog Island (58,640 ha), forming part of DBCA's "Return to 1616" project following the removal of exotic predators and herbivores [41]. Fifty-six individuals from both Bernier and Dorre Islands were proposed to be moved over 3 years (total $n = 112$). In this study, we aim to determine how the translocation history of *L. fasciatus* has affected the genetic health of all extant populations, particularly as serial translocations via intermediary captive populations have led to the possibility of genetic bottlenecks. We use genetic and population viability analysis models to explore the impact on source populations of concurrent harvesting scenarios and how the retention of genetic diversity can be maximised, and we predict growth rates and genetic diversity of the new Dirk Hartog Island and Mount Gibson populations. Most analyses are conducted under a baseline assumption of regular drought cycles that reduce reproductive success and survivorship, and we test the sensitivity of our results to both reduced and increased drought frequencies, as both scenarios are possible under climate change [42].

2. Materials and Methods

2.1. Tissue Samples

Ear biopsies were sampled over 19 years, from 1998 to 2017, from six locations: Dorre Island ($n = 79$), Bernier Island ($n = 51$), Faure Island ($n = 10$), the Peron CBC ($n = 73$), the Return to Dryandra Captive Facility (hereafter Dryandra, $n = 6$), and Wadderin Sanctuary ($n = 17$) (Figure 1d). All available *L. fasciatus* samples ($n = 236$, Spreadsheet S1) were analysed.

2.2. DNA Extraction and Nuclear Microsatellite Amplification

Genomic DNA was extracted from ear tissue using a standard "salting out" protocol [43]. Ten microsatellite loci were initially amplified using primers derived from the tammar wallaby (*Macropus eugenii*; Me14, Me17; [44]), yellow-footed rock-wallaby (*Petrogale xanthopus*; Y105, Y175, Y151, Y148; [45]), and allied rock-wallaby (*P. assimilis*; Pa593, Pa297, Pa385, Pa55; [46]). Briefly, for samples prior to 2016, polymerase chain reactions (PCRs) were carried out for individual microsatellite primers in a total volume of 30 µL with ~100 ng DNA, 1X PCR buffer, 400 µM of dNTPs, 2 mM $MgCl_2$, 0.2 µM of each primer, and 0.825 U Taq or, after 2016, in three PCR multiplexes using the Qiagen Multiplex PCR kit for a total volume of 7.5 µL per multiplex with ~5–10 ng DNA, 1X Qiagen buffer, 0.2 µM primer mix, and water (Table S1). Fluorescently labelled DNA fragments were separated using an ABI373xl capillary sequencer (Applied Biosystems, Foster City, CA, USA) and scored manually with the aid of GENEMARKER software (v1.5, Soft Genetics, State College, PA, USA). Allele size was determined by co-running a Genescan500 standard (Applied Biosystems, Melbourne, Australia). Data were checked for input errors and duplicate genotypes using the Excel Microsatellite Toolkit add-in [46]. Deviations from Hardy-Weinberg equilibrium and linkage disequilibria were tested using GENEPOP v4.1.4 [47]. For markers with fewer than five alleles, a complete enumeration algorithm was used to estimate the exact *p* value; and for markers with five or more alleles, the Markov chain algorithm of Guo and Thompson [48] was used to generate an unbiased estimate of the exact *p* value. The presence of any null alleles was tested with MICROCHECKER v2.2.3 [49].

2.3. Statistical Analyses of Genetic Data

The programme STRUCTURE v2.3.4 [50] was used to estimate the number of genetically distinct populations and to assign individuals to populations. STRUCTURE uses a Bayesian clustering method to assign individuals to one of k populations and to estimate the degree of inter-population admixture. While some assumptions made by the software will likely not be met by the island populations in our system, such as the distribution of genotypes under Hardy Weinberg Equilibrium (HWE) and marker linkage equilibria, this approach still provides a useful assessment of population genetic divergence. Our models assumed no admixture between locations, but since in some cases locations represent recently sub-sampled source populations, and therefore could have shared allele frequencies, we ran models that assumed either correlated or uncorrelated allele frequencies between groups for comparison. After preliminary assessment of convergence times for the Monte Carlo Markov chain, a burn-in period of 100,000 steps was chosen, followed by 1,000,000 steps of the chain. To estimate k, four replicate runs at each value of k from 1 to 8 were performed, and the most likely value was estimated from the plot of ln Pr (X|k) vs. k, as well as from Evanno's method [51] which plots Δ k (a second order rate of change of ln Pr (X|k)) vs. k, using CLUMPAK beta version [52]. STRUCTURE figures were generated using DISTRUCT v1.1 [53] in CLUMPAK [52]. To assess genetic similarity between populations using an approach that does not rely on the same biological assumptions, a principal coordinates analysis was conducted based on genetic distance, run in GENALEX v6.503 [54,55]. The level of genetic differentiation among all populations was determined by estimating pairwise F_{ST} and Jost's Dest, also in GENALEX v6.503. Probabilities were calculated as the proportion of times the observed data generated values greater than values generated from 999 random permutations.

Standard genetic diversity metrics, including mean number of individuals per marker, mean number of alleles per marker, and observed and expected heterozygosities, were estimated for each sampling site, in GENALEX v6.503. Allelic richness, a measure of allelic diversity that controls for variable sample size, was estimated in HP-RARE [56] based on rarefaction to ten individuals. As marker Me17 was monomorphic except for one individual in the Faure population, in which it was homozygous for the alternate allele, allelic richness was calculated with and without this marker. Population inbreeding coefficients were measured using Wright's F_{IS} in GENEPOP v4.1.4 [47]. Expected and observed heterozygosities, allelic richness, and FIS measurements were made for source populations separated into year cohorts. We estimated the genetic effective population size (Ne) using the linkage disequilibrium method on populations with a sample size of 25 or more [57–59], excluding singleton alleles (those that occur in one copy in one heterozygote) to prevent an upward bias in Ne estimation, as implemented in NeESTIMATOR v2.1 [60].

To assess whether any of the managed or natural populations underwent a genetic bottleneck, we compared the heterozygote distribution for each marker with the number of alleles for each population in BOTTLENECK v1.2 [61], using the two-phase substitution model with default proportions of the infinite alleles and single stepwise mutation models. A one-tailed Wilcoxon signed-rank test was used to estimate the likelihood that the observed data deviate from what is expected under mutation-drift equilibrium, the most appropriate test with fewer than 20 loci [62,63]. We also reported whether the allele frequency distribution deviated from an L-shaped mode, which can be a qualitative assessment of whether a population has passed through a genetic bottleneck [64].

Queller and Goodnight's [65] estimator was used to measure the mean pairwise relatedness within each population, as an indicator of genotype diversity and identity by descent within source populations. Whether the mean pairwise relatedness of a population was statistically different from 0 (defined as mean pairwise relatedness across the entire data set) was estimated by randomly permuting the data in the population pairwise matrix 999 times, and determining the proportion of permutations that gave a relatedness value greater than the observed value, as calculated in GENALEX v6.503.

2.4. Population Modelling

To estimate the number of founders needed to retain low-frequency alleles within newly established populations, scenarios were simulated in ALLELERETAIN v1.1 [66,67] implemented in R v3.6.2 [68]. ALLELERETAIN is an individual-based model that simulates population growth using a user-defined suite of parameters based on life-history traits. It estimates the probability of retention of selectively neutral alleles in founder populations over generations, with the starting frequency of these alleles set by the user. We were interested in assessing the impacts of translocating individuals of varying founder group size (N) on allele retention. To do this, we tested founder N values from 20 to 200 in increments of 20 individuals with allele frequency set to 0.05. We then repeated each scenario but considered the translocation to have occurred during a drought year. This was simulated by setting the initial survival after translocation to 0.4 and excluding reproduction for the first breeding cycle. For all simulations, we assumed that one *L. fasciatus* breeding cycle lasts 9 months [69]. Details and justification of the demographic parameters used in the model are provided in Table 1 and Table S2.

Table 1. Life history parameters used for *L. fasciatus* population modelling in VORTEX and ALLELERETAIN. Parameter values were obtained from data in [33,34,41,70] and refined by C.S. and J.S. Where a parameter is not listed, defaults are used. V: VORTEX; A: ALLELERETAIN; EV: environmental variation; BHW: banded hare wallaby; DDR: density-dependent reproduction function.

Life History Parameter	Value	Used in
Species description		
Inbreeding depression		V
Lethal equivalents	3.14	V
% due to recessive lethal	50	V
EV concordance of reproduction and survival	0.5	V
EV correlation among populations	0.5–0.8	V
Reproductive system		
Reproductive system	polygynous	V/A
Duration of breeding cycle in days	274	V/A
Age of first offspring for females/males	9 months/18 months	V/A
Maximum age of reproduction	8 years	V/A
Maximum lifespan	10 years	V/A
Maximum number of broods per breeding cycle	1	V/A
Maximum number of progeny per brood	1	V/A
Mean female number of progeny per breeding cycle	1	A
Mean male lifetime reproductive success (±SD)	7 ± 3	A
Sex ratio at birth	50	V/A
Reproductive rates		
% adult females breeding	90% with DDR	V
EV in % breeding	18	V
Distribution of broods per breeding cycle		
0 broods	0	V
1 brood	100%	V
Number of offspring per female brood		
1 offspring	100%	V

Table 1. *Cont.*

Life History Parameter	Value	Used in
Mate monopolisation		
% males in breeding pool	85	V
% males successfully siring offspring	63	V
Mortality rates		
Females		
Mortality age 0 to 1 (±SD)	40 (±10)	V/A
Annual mortality after age 1 (±SD)	10 (±3)	V/A
Males		
Mortality age 0 to 1 (±SD)	40 (±10)	V/A
Annual mortality after age 1 (±SD)	10 (±3)	V/A
Catastrophes		
Number of types of catastrophes	1 (drought)	V
Frequency	1 in 6.25 calendar years[a]	V
Severity	50% reduction in survival and reproduction	V
Initial population size		
Bernier	2000	V/A
Dorre	2000	V/A
Faure	300	V/A
Carrying capacity, K (SD due to EV)		
Bernier	3000 (300)	V/A
Dorre	3000 (300)	V/A
Faure	1000 (100) or 3000 (300)	V/A
DHI	(1000)	V/A
Mt. Gibson	5000 (500)	V/A
Genetic management		
Number of neutral loci to be modelled	7 empirical, 1 simulated	V
Initial minor allele frequency	0.05	A
Scenario settings		
No. replicates	1000/100	V/A
No. years	50 calendar years	V/A

To estimate the impact of current translocation programmes on extinction probabilities and heterozygosities of the newly established *L. fasciatus* populations and the source populations, population viability analysis (PVA) simulations were run in VORTEX v10.17.2 [71,72]. VORTEX is an individual-based model that uses Monte Carlo simulations to estimate how factors intrinsic to individuals within populations alter growth rates, birth rates and extinction probabilities [72]. Initially, we developed a baseline PVA for a closed population with a starting N of 2000 individuals and a carrying capacity of 3000, where droughts at an average frequency of 1 in 6.25 calendar years reduced survival and reproduction by 50%, similar to empirical observations [33,34]. Although there is no empirical evidence for inbreeding depression, we included it in our models as there has been as yet no formal effort to detect it and populations do regularly pass through extreme population bottlenecks [33,34]. However, as populations appear to persist through these bottlenecks without obvious signs of depression, we halved the default number of lethal equivalents. Other island populations of marsupials are known to survive small effective population sizes and high levels of inbreeding [73]. Baseline model parameters are provided in Table 1 and Table S3.

Various translocation scenarios (each of 1000 replicates) were then simulated that tested the number of founders and source populations needed to maximise genetic diversity and population growth for a translocation programme that, due to the logistical cost to the management agency, runs for a maximum of two years (Table 2). Bernier Island was chosen for single source population translocations for two reasons. Firstly, as this population is smaller in size and has lower genetic diversity, it represents a conservative, worst case scenario, and secondly, this island is easier to access due to protected landing beaches. To quantify the impact of drought on translocation success, the best performing of these scenarios was re-run with both increased and decreased drought frequencies, and with no drought. Finally, a case study (scenario 8) was simulated based on the recent movements of *L. fasciatus* from Bernier and Dorre Islands to Dirk Hartog Island, and from Bernier, Dorre, and Faure Islands to the Mount Gibson Wildlife Sanctuary. Although the carrying capacity of Faure Island could theoretically be similar to Bernier and Dorre Islands (predicted here to be 3000) based on land area, there is currently no empirical evidence for this and the census population size remains well below 1000. We therefore tested two different carrying capacities for Faure Island, 3000 and a more conservative 1000. To further assess the impact of drought cycles in translocation planning, both the current and post-drought estimates of source population sizes were used as inputs in case study models.

Table 2. Hypothetical and case study translocation scenarios * examined using population viability analyses.

Target Population	Scenario	Description
Dirk Hartog Island	1	One translocation from Bernier Island in year 1
	2	Individuals translocated from Bernier Island only, half in each of first two years
	3	Individuals translocated from Bernier and Dorre Islands, half from one island in year 1 and half from other island in year 2
	4	Individuals translocated from Bernier and Dorre Islands, half from each island in year 1
	5–7	Best performing scenario from 1 to 4, with drought frequencies of no drought, 1 in 10 years, and 1 in 5 years
Dirk Hartog Islandand Mount Gibson Wildlife Sanctuary	8	To Dirk Hartog Island: six individuals from Bernier and six from Dorre in BHW year 1; 50 from Bernier in BHW year 2; 50 from Dorre in BHW year 3To Mount Gibson: 23 individuals from Bernier, 39 from Dorre, and 10 from Faure in BHW year 1; 37 from Bernier, one from Dorre, and 20 from Faure in BHW year 3

* Different founder numbers were trialed for scenarios 1–4, ranging from 60 to 140 every 20 individuals. Scenario 8 is a case study representing current translocations. Unless otherwise stated, a drought is simulated at a frequency of 1 in 6.25 years.

3. Results

3.1. Marker Performance

Of the ten microsatellite markers amplified, three were monomorphic in this species. The informativeness of the seven polymorphic markers varied, with the number of alleles ranging from two (Y175 and Me17) to eight (Pa593) (Table S4). Expected heterozygosity (mean ± SE) ranged from 0.03 ± 0.03 (Me17), which was polymorphic in only one population, to 0.66 ± 0.02 (Pa593). The mean percentage of samples with missing genotype data across all populations ranged from 0.0% (markers Pa593, Me14 and Y105) to 5.1% (marker Y148). Marker Y175 violated the Hardy–Weinberg equilibrium ($p < 0.05$) in three populations out of six. Analysis in MICROCHECKER suggested the unexpectedly high homozygote frequency was caused by a null allele in all three populations. The percentage of samples with missing genotype data for this marker ranged from 0.0% (Dryandra) to 12.3% (Peron CBC), with a mean of 3.8%. After assessing the impact of null alleles on the estimation of global and pairwise FSTs in FreeNA [74], no significant differences were observed between estimates calculated with and without correcting for null alleles. Considering this, and the low number of markers, Y175 was retained for further analysis.

3.2. Genetic Diversity

Structure analysis revealed that the most likely number of genetically distinct clusters of *L. fasciatus* is two (k = 2; Figure 1e), representing the two remnant wild populations on Bernier and Dorre Islands. There was very good agreement in results whether allele frequencies were assumed to be correlated or uncorrelated between locations, and the absolute probability for k = 2 was slightly greater for correlated allele frequencies. Of the contemporary translocated populations, Faure Island and Wadderin are predominantly of Bernier Island origin, with negligible genomic contribution from Dorre Island. Of the extinct captive populations, Peron CBC was an extension of Bernier Island, whereas Dryandra was predominantly of Dorre Island origin. A small amount of potential admixture was detected between Bernier and Dorre Islands. In total there are ten individuals (four from Bernier and six from Dorre) that have an 80% or more probability of assignment to the other source population. Principal coordinate analysis agreed with the STRUCTURE results (Figure 1f). Both pairwise FST and Jost's Dest measurements were similar (F_{ST} = 0.026, $p < 0.001$; Dest = 0.029, $p < 0.001$), reflective of weak, significant divergence between the two parental populations. A summary of all pairwise FSTs is provided in Table S5, which shows that greatest distance exists between Bernier Island-derived populations and Dryandra (the sole Dorre Island-derived population). Pairwise relatedness analysis revealed that one source population (Bernier Island, $p < 0.005$), one extinct captive population (Peron CBC, $p < 0.002$), and one contemporary translocated population (Wadderin, $p < 0.014$) have relatedness values statistically greater than zero (Figure S1).

Overall, genetic diversity was low across all populations (mean expected heterozygosity, HE = 0.34–0.45). Of the source populations, Dorre Island exhibited higher heterozygosity and lower inbreeding than Bernier Island, although allelic richness was comparable. Dorre Island had one private allele (alleles present in one population only) for marker Y105 (q = 0.02), Faure Island had two (Pa593, q = 0.05; Me17, q = 0.10), and Peron CBC had one (Y148, q = 0.01). When the two parental populations were compared only to each other, Dorre Island had four private alleles and Bernier Island had one. Diversity estimates within year cohorts for both Bernier and Dorre Islands showed a similar trend of higher than expected heterozygosities in the late 1990s, and lower than expected heterozygosities in more recent cohorts. Both expected heterozygosity and allelic richness have declined over the sampling period (Supplementary Figure S2). There were no significant differences in genetic variation between translocated and source populations except for Faure Island which showed a significant decrease in observed heterozygosity compared to Peron CBC using a one-tailed Wilcoxon signed-rank test ($p < 0.05$), and this island also indicated significant inbreeding due to non-random mating. Considering the two parental populations as one remnant *L. fasciatus* metapopulation, Faure has retained 93% of allelic diversity, while Wadderin has retained 87%. Calculating allelic richness without marker Me17 increases its value in all populations, but Faure no longer has the highest allelic diversity (Table S6). Caution is needed when interpreting all diversity results, however, as sample sizes for diversity measurements were often very low and varied between sites and temporal data.

Estimates of effective population sizes in extant populations reflected overall diversity measurements (Table 3), as Dorre Island had the highest Ne in the 2016–2017 cohort (Ne = 140, 95% CI 29-∞), which was 71% higher than for Bernier Island *L. fasciatus* sampled in the same period (Ne = 82, 95% CI 12-∞). Values of infinity for upper confidence limits arise from a lack of evidence for variation in the genetic characteristic which, in this case, is likely due to the small, homogeneous marker panel [59,60]. Our estimates of effective population size should therefore be considered cautiously and await confirmation with a more powerful dataset.

Table 3. Population-level diversity metrics using seven polymorphic microsatellite markers.

Population	N [a]	HWE [b]	NA	PA		AR (±s.e.)	HE (±s.e.)	HO (±s.e.)	F_{IS} (±s.e.)	Ne (95% CI)
Bernier Island (all)	51	1/6	20	0		2.47 (0.15)	0.36 (0.09)	0.30 (0.08)	0.14 (0.05)	
Bernier (1998)	6					n/a	0.38 (0.11)	0.43 (0.15)	−0.06 (0.17)	n/a
Bernier (2010–2011)	9					n/a	0.34 (0.10)	0.23 (0.07)	0.31 (0.10)	n/a
Bernier (2016–2017)	33					2.40 (0.15)	0.36 (0.08)	0.31 (0.07)	0.11 (0.04)	82 (12, ∞)
Dorre Island (all)	79	2/6	23	1		2.46 (0.10)	0.42 (0.08)	0.41 (0.08)	0.03 (0.07)	
Dorre (1995–1996)	7					n/a	0.44 (0.08)	0.61 (0.11)	−0.39 (0.07)	n/a
Dorre (1999–2000)	8					n/a	0.45 (0.09)	0.45 (0.09)	0.00 (0.07)	n/a
Dorre (2013)	11					2.41 (0.29)	0.42 (0.08)	0.43 (0.09)	0.01 (0.08)	n/a
Dorre (2016–2017)	52					2.40 (0.11)	0.40 (0.07)	0.38 (0.08)	0.07 (0.10)	140 (29, ∞) [c]
Faure Island (2017)	10	0/6	18	2		2.57 (0.40)	0.39 (0.08)	0.29 (0.09)	0.36 (0.14)	n/a
Wadderin (2018)	17	1/5	16	0		2.27 (0.27)	0.36 (0.10)	0.34 (0.13)	0.09 (0.20)	n/a
Peron CBC (2006–2013)	73	1/5	19	1		2.41 (0.15)	0.36 (0.10)	0.33 (0.10)	0.08 (0.08)	20 (7, 57)
Dryandra (1999–2002)	6	0/6	18	0		n/a	0.40 (0.08)	0.50 (0.12)	−0.28 (0.15)	n/a
	236				μ (±s.d.)	2.34 (0.13)	0.39 (0.02)	0.37 (0.08)	0.07 (0.21)	

N: number of samples; HWE: Hardy–Weinberg equilibrium; NA: total number of alleles; PA: private alleles; AR: allelic richness (rarefied to 10 individuals); HE: expected heterozygosity; HO: observed heterozygosity; F_{IS}: inbreeding coefficient; Ne: effective population size. [a] Diversity estimates for samples of size 10 or fewer should be treated with caution. [b] Proportion of polymorphic markers that violate HWE at $p < 0.05$; [c] value obtained only with no critical minor allele frequency threshold. Heterozygosity metrics and inbreeding coefficients are also calculated for different time periods for Bernier and Dorre Islands, when at least five individuals were sampled for that time period.

Combining samples from all years, only one source population, Dorre Island, had statistical support (p = 0.039) for passing through a genetic bottleneck (Table 4). However, when Bernier and Dorre Islands were separated into year cohorts, both populations showed evidence of genetic bottlenecks in the earlier cohorts (1990s), and for Dorre Island also in 2013. Of the managed populations, Wadderin was the only population showing evidence of a genetic bottleneck (Table 4). Thirty individuals and 10 polymorphic loci are recommended to achieve power of at least 0.8 with the Wilcoxon signed-rank test however, so these results should be considered with caution. Both Wadderin and Peron CBC had an allele frequency distribution shifted towards more common alleles, consistent with a bottleneck, although this qualitative test requires at least 30 individuals and 10 to 20 polymorphic loci to be reliable [61].

Table 4. Results of two tests for a genetic bottleneck of source and translocated populations.

Population (Time Period)	n	Bottleneck Test	
		Wilcoxon (One-Tailed, H Excess)	Mode Shift
Bernier Island (all)	51	0.281	No
Bernier Island (2016/2017)	33	0.500	No
Bernier Island (2010/2011)	9	0.109	Yes
Bernier Island (1998)	6	0.016	Yes
Dorre Island (all)	79	0.039	No
Dorre Island (2016/2017)	52	0.422	No
Dorre Island (2013)	11	0.008	Yes
Dorre Island (1999/2000)	8	0.008	Yes
Dorre Island (1995/1996)	7	0.016	Yes
Peron CBC (from Bernier; t = 25, y = 1998)	73	0.219	Yes
Dryandra (from Dorre and Peron CBC; t = 25, y = 1998)	6	0.578	No
Faure Island (from Peron CBC; t = 91, y = 2004 to 2013)	10	0.281	No
Wadderin (from Peron CBC; t = 12, y = 2013)	17	0.031	Yes

Integers represent p values, with significant values in bold. n: sample size; t: translocated population size; y: year of translocation. The two-phase mutation model of microsatellite evolution is assumed with a 70:30% ratio of single stepwise mutation to infinite allele models. Year cohorts with fewer than five individuals were omitted from analyses. Estimates based on sample sizes of fewer than 30 should be considered tentative.

3.3. Modelling Conservation Translocations

In the absence of periodic droughts, ALLELERETAIN demonstrated the strongest increases in the probability of retaining low-frequency alleles when the founder population size approached 20, and began to plateau after 40 (Figure 2). Sixty founders are needed to retain 90% of alleles at a frequency of 0.05, and 80 founders are needed to retain 95% of alleles (recommended thresholds) at a frequency of 0.05. These numbers rose to 120 and 140 founders, respectively, when realistic impacts of drought were simulated.

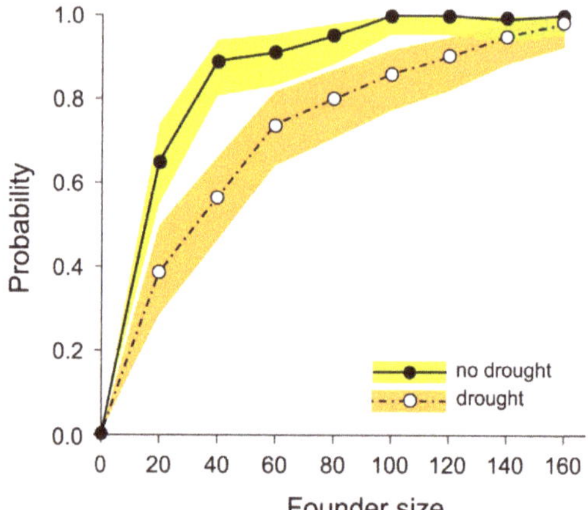

Figure 2. Probability of retaining a selectively neutral allele at a starting population frequency of 0.05 after 50 calendar years for translocated *L. fasciatus* populations with various founder sizes, with and without impacts of drought. Shaded ribbons represent 95% confidence intervals.

Population viability analysis modelling showed that when contrasting founder population sizes between 60 and 140, survival probabilities increase with increasing founder number until this relationship begins to asymptote from around 100 individuals (Table S7a–d). This level of harvesting also had limited impact on the source populations. Mixing the two source populations of Bernier and Dorre Islands increased heterozygosity in the translocated population, relative to only using founders from Bernier Island, and there was minimal impact of moving all individuals in one year versus splitting the translocation over two years (Table 5). Considering the additional logistical cost of a second year of translocation activities, the best scenario in terms of maximising survival probability, genetic diversity (heterozygosity and allelic diversity), and the final population size of the translocated population was to harvest the two source populations in one year (Scenario 4, Table 5).

Table 5. Viability of a translocated *L. fasciatus* population after 50 calendar years, testing four scenarios. One hundred individuals translocated in one or over two years, including a drought with a mean frequency of one every 6.25 calendar years. Carrying capacity was set at 10,000 and scenarios were run for 1000 replicates.

Scenario	Description	P (surv)	HE	N
1	100 in year 1 from Bernier Island	0.76	0.311	2094
2	50 from Bernier Island in year 1 and year 2	0.79	0.330	2489
3	50 from Bernier Island in year 1, 50 from Dorre Island in year 2	0.80	0.368	2251
4	50 from Bernier Island, 50 from Dorre Island in year 1	0.79	0.362	2277

P (surv): mean probability of survival; HE = mean expected heterozygosity using empirical estimates of allele frequencies to determine starting heterozygosities; N: final mean population size of extant populations.

Simulations that varied drought frequency indicated the likelihood of an appreciable impact on the survival probability of a translocated population after 50 calendar years, dropping from 100% under the assumption of no drought, to 79% with a realistic drought frequency of 1 in 6.25 years, and 58% with a drought frequency of 1 in 5 years (Figure 3). There was also predicted to be a profound effect on the final N, as predicted carrying capacity (set to 10,000) was reached with no drought, whereas final N was estimated at 2277 individuals with a 1 in 6.25-year drought and 1064 at 1 in 5 years (Figure 3).

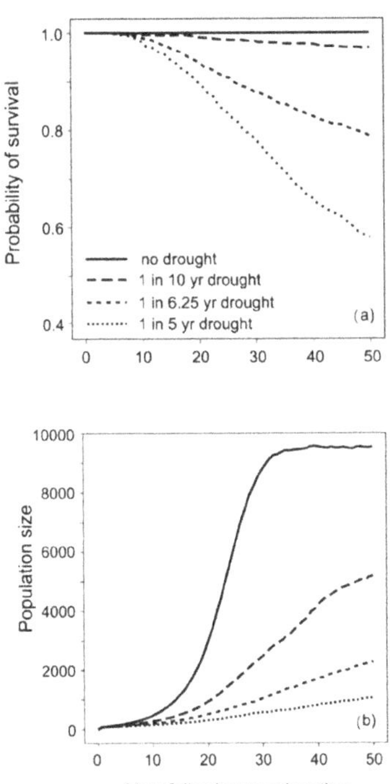

Figure 3. Performance over 50 years for 100 translocated *L. fasciatus* to DHI with drought frequencies of 1 in 5 years, 1 in 6.25 years, 1 in 10 years, or no drought (a) Survival probability. (b) Population size.

Simulation of translocations recently conducted by management agencies between 2017 and 2018—harvesting Bernier and Dorre Islands for translocation to Dirk Hartog Island, and concurrent harvesting of Bernier, Dorre, and Faure Islands for translocation to Mount Gibson—including the impacts of 1 in 6.25-year droughts, predicted good survival probabilities over 50 years for the translocated populations of 83% and 84%, respectively. This result was also achieved even if harvesting occurred with a reduction in the size of source populations following drought (Table 6). In contrast, the timing of harvesting had a more significant effect on the source populations. Harvesting at current census population sizes (Nc) resulted in high survival probabilities (99%, 98%, and 93% for Bernier, Dorre, and Faure Islands, respectively; Table 6), assuming Faure Island to have a carrying capacity (K) of 1000. Bernier and Dorre Island population sizes stabilise at ~82% of current estimates, whereas the Faure Island population stabilises after ~65% growth from current estimates to a census size of around 500. If harvesting were to occur immediately after a drought, predicted survival probabilities reduced after 50 calendar years, most markedly to 60% on Faure Island, and there was a further reduction in predicted Nc in Bernier and Dorre Island populations to around 74% of current estimates, whereas growth on Faure Island was limited to 29% (Nc ~386). Interestingly, when Faure Island was assumed to have a K of 3000, there was little improvement in survival probability post-harvest if a drought census size was assumed (63% with K of 3000 vs. 60% with K of 1000). However, there was a dramatic increase in predicted Nc as 1400 was reached when harvesting occurred with a non-drought Nc and 900 with a drought impacted Nc, compared to 495 and 386 with a K of 1000, respectively.

Table 6. Comparison of population viability analysis (PVA) outputs after 50 calendar years when harvesting source populations using either current conservative source census sizes, or likely source census sizes following droughts. Models are based on scenario 8, describing recent and ongoing movement of *L. fasciatus*, and assuming an average drought frequency of 1 in 6.25 years.

	Conservative Current Census Sizes of Source Populations $N_{Bernier} = 2000$, $N_{Dorre} = 2000$, $N_{Faure} = 300$	Census Sizes Following Drought $N_{Bernier} = 500$, $N_{Dorre} = 500$, $N_{Faure} = 75$
(a) Target Populations		
Dirk Hartog Island		
P (surv)	0.83	0.81
N	2363	2436
HE	0.367	0.367
Mount Gibson		
P (surv)	0.84	0.84
N	1600	1565
HE	0.372	0.372
(b) Source Populations		
Bernier Island		
P (surv)	0.99	0.94
N	1640	1474
HE	0.359	0.353
Dorre Island		
P (surv)	0.98	0.95
N	1647	1465
HE	0.417	0.408
Faure Island *		
P (surv)	0.93/0.93	0.60/0.63
N	495/1401	386/901
HE	0.369/0.374	0.344/0.343

P (surv): mean probability of survival; N: final mean population size of extant populations; HE = mean expected heterozygosity. * Numbers to the left of the slash are for a carrying capacity of 1000, and to the right are for a carrying capacity of 3000. Scenarios were run for 1000 replicates.

4. Discussion

As the remnant range of *L. fasciatus* consists of just two adjacent offshore islands, the species-level impact of detrimental stochastic events affecting one or both islands could be profound. Establishing insurance populations and restoring the species to other suitable habitats free from introduced predators is a fundamental management objective, as is the retention of remnant genetic diversity to maintain adaptive capacity. While it is difficult to make conclusive summaries from the limited genetic data we provide here, we nonetheless attempt to explain observed trends that justify a more extensive genetic study. For example, our data suggest that both remnant island populations underwent genetic bottlenecks in the mid to late 1990s, potentially linked to the boom and bust cycles of *L. fasciatus* that occur in response to periodic droughts and thus a marked reduction in primary productivity [33,34]. Further, we show that the impact of past conservation actions (captive breeding and translocations) has possibly manifested in different ways in the two surviving translocated populations (Faure Island and Wadderin). Based on the limited samples available, Faure Island may be inbred due to non-random breeding but has relatively high allelic diversity, whereas Wadderin does not show signs of genetic inbreeding due to non-random breeding but may have passed through two selectively stronger bottlenecks (Bernier Island to Peron CBC, and Peron CBC to Wadderin). Our population viability and genetic modelling has revealed the importance of considering periodic fluctuations in population size when planning translocations and informed on the number of founders needed to avoid genetic bottlenecks. Taken together, these lines of evidence point to a critical need for genetic management and guidance for future translocations.

4.1. Genetic Diversity in Remnant Populations

Island populations are expected to lose genetic diversity as a result of genetic drift, especially in the absence of migration (i.e., gene flow) that acts to increase allelic diversity [75,76]. Here, we found that genetic diversity (allelic richness and heterozygosity) of *L. fasciatus* on Bernier and Dorre Islands was approximately two-fold lower relative to values reported for mainland populations of other Australian marsupials [77,78], and reviewed in [79], which is unsurprising considering the low number of polymorphic markers used in this study. Interestingly, values were similar to the Shark Bay island population of another hare-wallaby, the rufous hare-wallaby (*Lagorchestes hirsutus*), which was measured with shared markers and also had lower diversity than its mainland counterpart [80]. We also identified extremely low effective population sizes for *L. fasciatus* (Table 3) given census sizes of ~2800 for Bernier Island and ~2500 for Dorre Island [32,33], leading to good agreement in Ne/Nc ratios of 0.05 and 0.07, respectively. These estimates are several times smaller than those for other mammals, e.g., 0.2 to 0.8 reported in [81,82] and a mean of 0.75 reported in [83]. However, due to our low power to estimate Ne from genetic data, these trends require confirmation.

Bernier and Dorre Islands have been separated from mainland Australia for 8000 years (around 4000 generations), and from each other for around 5000 years [84,85], providing substantial opportunity for stochastic loss of allelic variation through genetic drift. The finding that around 10 remnant source population individuals (out of 130) were assigned with 80% or more probability to the other source population is therefore surprising. This observation could be consistent with recent gene flow, leading to migrant populations that have remained relatively isolated from the resident populations. However, this is unlikely, especially as it would have needed to have happened reciprocally. A more likely explanation is either a lack of discriminatory power in our markers, perhaps due to being developed in sister species [86], or there has been a technical issue, such as mis-labelling. Adequate sampling to properly determine genetic structure and the use of more powerful markers (e.g., single nucleotide polymorphic markers generated through next-generation sequencing), would resolve these alternative hypotheses.

4.2. Impact of Past Translocations on Genetic Diversity

The expected heterozygosities of either of the established translocated populations were not less than their source population (Peron CBC) or the parental population (Bernier Island). Despite describing two examples of serial translocations occurring from a single source population via a captive breeding programme, and hence providing opportunity for two bottleneck events, in only one case did we find evidence for a bottleneck—the Wadderin population translocated in 2013 with 12 founders. Wadderin was founded from the Peron CBC population, which itself was founded from Bernier Island (Figure 1d), and each translocation contributed a 2.4% and 5.7% loss of allelic diversity, culminating in a total loss from the Bernier source population of 8.1%. Faure, on the other hand, did not undergo a genetic bottleneck according to our analysis, and had a relatively high allelic diversity. The high allelic diversity is surprising considering its demographic history, and we found it to be driven by the discovery of an additional allele (marker Me17), present in one individual in homozygous form. While this allele could have arisen in the translocated Faure Island population, its independent verification would benefit from expanding sampling of Faure and Bernier Islands. While we were not able to detect a genetic signal of a bottleneck in the Peron CBC population, founded by 25 individuals, this does not preclude the possibility that a bottleneck has occurred that we were not able to detect or from which the population later recovered.

Since translocations began to Faure Island, however, observed heterozygosity appears to have decreased below equilibrium expectations and inbreeding coefficients have increased, possibly due to limited dispersal and increased breeding between related animals post-translocation. Another possibility is an increase in mating success heterogeneity [78]. These factors could also have contributed to the island's notably low effective population size estimate. This is of concern, as small populations are prone to higher rates of loss of genetic variation via genetic drift, especially if inbreeding depression leads to lower rates of reproduction and further reductions in population size [17]. However, the apparently high inbreeding does not appear to be associated with poor health of the Faure Island population, as it is well established with ongoing recruitment [87]. One explanation for this apparent paradox is that the boom and bust population cycles characteristic of Shark Bay *L. fasciatus* populations have led to a purging of lethal alleles, negating any detrimental effect of inbreeding [34]. A more prosaic explanation is that the Faure Island sampling cohort was both small ($n = 10$) and consists of a disproportionately high number of related individuals, leading to an overestimate of the inbreeding coefficient. A more comprehensive genetic analysis and a future study of the social dynamics of the Faure Island population could resolve this uncertainty.

While our bottleneck results require independent replication, a lack of bottleneck signal for the Peron CBC population with 25 founders supports results from our allele retention modelling. We show that a 70% probability of retaining a low frequency allele is maintained with a founder size of 25, but that retention probabilities drop dramatically with founder sizes less than 20. Other populations that have passed through bottlenecks tend to have greater founder numbers. For example, in birds, a bottleneck occurred when the Hawaiian gallinule (*Gallinula galeata sandvicensis*) was reduced to approximately 60 individuals [88], and reductions to 100 individuals may be sufficient to observe bottleneck effects in other avian species [89]. In the Maud Island frog (*Leiopelma pakeka*), simulations indicate that a sustained increase in irreversible allele loss begins when populations are reduced to 140 [90]. The apparent low power to detect genetic bottlenecks in translocated *L. fasciatus* populations could reflect relatively small changes in allelic diversity between source and founder populations (due to low effective population sizes and genetic variation in the source), compounded by our small panel of only moderately polymorphic markers. Further, the relatively rapid demographic recovery inherent to *L. fasciatus* during the boom phase of natural population cycles could mask any bottleneck signal that depends on heterozygosity excess [62]. Therefore, ongoing monitoring, including genetic monitoring, of translocated populations during their establishment would provide useful data with which to detect any detrimental genetic effects that might be initially masked.

4.3. Towards an Optimal Translocation Protocol for L. fasciatus

A clear trade-off exists in conservation translocation programmes between maximising viability of a new population and minimising the negative impact on critical source populations [90]. For example, sufficient founders are required to ensure a translocated population is buffered from some post-establishment mortality, as well as to retain rare alleles that bolster evolutionary potential [67,90]. Smaller founder numbers, on the other hand, reduce any negative impacts on source populations, but translocated populations will be more sensitive to mortality and the loss of rare alleles. Our PVA models for *L. fasciatus* show that for newly translocated populations, survival probabilities increase with increasing founder number until this relationship asymptotes at around 120 individuals, and the release of 100 individuals ensures good growth and survival in a hypothetical haven (island or fenced sanctuary with a carrying capacity of 10,000). That carrying capacity did not significantly impact 50-year predicted survival probabilities for the Faure Island population when using reduced drought-influenced census sizes, indicates the sensitivity of the successful establishment of *L. fasciatus* translocated populations to the founder number, and suggests that sufficient numbers are required to withstand regular population reduction caused by stochastic events. Importantly, harvesting the numbers recommended here has no apparent impact on the critical source populations, either in terms of population growth or survival probability over a 50-year period.

Our allele retention models predict that 80 founders are needed to retain 95% of low-frequency alleles, noting that retaining 90 or 95% genetic variation are recommended minimum thresholds for maintaining the evolutionary potential of populations [17,91]. Hence, aggregating the results of both modelling processes suggests that at least 100 individual founder animals are needed to maximise viability and retention of allelic diversity, a comparable figure to other taxa (e.g., 120 in frogs [90], 60 to 120 in passerines [19]). Further, by simulating the effects of periodic drought on population growth and survival, we have provided predictions about when best to translocate. For example, if harvesting were to occur immediately after a drought, we predict a 10–22% reduction in the census size of the source population after 50 years, relative to harvesting when populations are at their peak. In turn, we predicted that 140 founders would be required to retain 95% of alleles with a population frequency of 0.05 in a new population under drought conditions (as opposed to 80 founders under no drought); a 75% increase in the number harvested from the critical source populations. Therefore, climatic cycles appear to be an important consideration in translocation programmes, and we advise that the movement of animals should occur outside periods of low rainfall and when populations have recovered demographically from their impacts.

One way to increase genetic diversity of species that exist as only small and genetically depauperate remnants is to manage the species as a metapopulation [92]. However, whether to combine genetically distinct populations is contentious [1,7,78,93–95]. While there is an immediate and obvious benefit of maximising genetic diversity and heterozygosity, thereby reducing the chance of inbreeding depression, mixing comes with risks, including outbreeding depression [1]. For example, if source populations derive from different habitats or climates, and have been long isolated, local adaptation can cause the rise of locally adapted alleles and for gene complexes to be in gametic disequilibrium. Locally adapted alleles can be diluted and gametic associations can be broken by recombination after mixing between differently adapted individuals, resulting in hybrid individuals with low fitness [96–99]. Other risks include competition for resources when there are morphological and size differences between source populations, and behavioural differences could affect mate choice, potentially reducing Ne and leading to a bias in representation of future generations (e.g., as occurs in the boodie *Bettongia lesueur*, [78]). Here, however, the habitats of the two adjacent *L. fasciatus* source populations have been similar across the 4000–5000 years (approx. 2000 generations) of their separation [79]. Further, there are no known differences in size, morphology, or mating behavior between islands, suggesting a low probability of outbreeding depression [95]. Considering the overall low genetic diversity maintained in this species, and by mixing we achieve a 10–12% increase in HE of the recipient population compared to using only Bernier Island as a source population, we advocate mixing the two source island populations in future translocations and predict a low risk of any adverse consequences.

5. Conclusions

Our work shows how an integrated analysis of genetics and population modelling can be used to inform management planning by simulating sensitive trade-offs involved in translocating threatened species. While much work has been done on the numbers of founders needed to establish new populations, less is known about the impact of harvesting for translocations on source populations [89], which is highly relevant for species with few remnant populations remaining. We suggest that the optimal translocation protocol for *L. fasciatus* is to mix the Bernier and Dorre Island populations in the same year, harvesting 60 individuals from each island source population (total *n* = 120). This gives a good probability of retaining low-frequency private alleles from each island, maximises survival probability and heterozygosity of translocated populations, limits logistical cost, and, crucially, has no major impact on the source populations. As the translocations currently underway to Dirk Hartog Island and Mt. Gibson approximate these recommendations, it will be valuable to monitor these populations over time to compare performance to model predictions. It is also worth considering Faure Island as an alternative source population to Bernier Island for future translocations, as was done with the Mt. Gibson translocation, to relieve pressure on one of the two remnant wild populations. However, in addition to possible inbreeding, our modelling shows Faure Island is more sensitive to stochastic events due to a smaller census population, further compounded if its carrying capacity is less than Bernier and Dorre Islands, and so the PVA models developed here should be used to assess the impact of any future harvesting. Wherever feasible, and after consideration of potential over-harvesting, future supplementations of the established translocated populations on Faure Island and Wadderin Sanctuary should derive from Dorre Island which, as a genetic mixing event, would be expected to increase gene diversity of the recipient population. Further, Dorre Island has a higher heterozygosity, more private alleles and lower inbreeding than Bernier Island. Finally, while we show that *L. fasciatus* seems resilient to harvesting and is well suited to translocation programmes, including multiple harvests from source populations, regular droughts, and limited carrying capacity, which have a substantial impact on population viability, particularly for populations with census sizes less than 500. We recommend that translocations are avoided during extended periods of drought and subsequent demographic recovery, due to a notably lower growth rate of the source populations after harvesting, and reduced capacity of new populations to retain allelic diversity.

Supplementary Materials: The following are available online at http://www.mdpi.com/1424-2818/12/12/448/s1, Spreadsheet S1: Genetic samples; Table S1: PCR multiplexes; Table S2: Details and justification of the demographic parameters used in ALLELERETAIN; Table S3: Details and justification of the demographic parameters used in VORTEX; Table S4: Performance of seven polymorphic microsatellite markers; Table S5: Population pairwise FSTs and probability values; Table S6: Comparing allelic richness with and without marker Me17; Table S7a–d: Summary of PVA outputs for translocation scenarios; Figure S1: Pairwise relatedness; Figure S2: Temporal change in genetic diversity for the two remnant source populations.

Author Contributions: Conceptualization, N.J.M., K.O. and D.J.W.; formal analysis, D.J.W.; investigation, D.J.W. and K.O.; resources, K.O., P.B.S.S., M.S., C.S. and J.S.; data curation, K.O. and P.B.S.S.; writing—original draft preparation, D.J.W.; writing—review and editing, N.J.M., K.O., J.S., P.B.S.S., M.S., C.S. and D.J.W.; visualization, N.J.M. and D.J.W.; project administration, D.J.W. and N.J.M.; funding acquisition, N.J.M. and K.O. All authors have read and agreed to the published version of the manuscript.

Funding: This research was supported by the Australian Government's National Environmental Science Program through the Threatened Species Recovery Hub.

Acknowledgments: We thank Keith Morris and Manda Page for comments on earlier versions of the manuscript, John Kanowski for valuable comments on late versions of the manuscript, Mark Eldridge, Anthony Friend, and Mia Hillyer for help with conceptualisation, Saul Cowen for discussion on translocation strategies, Laura Ruykus and Kim Branch for collection of samples, and Shelley McArthur for assistance in the laboratory. The authors would also like to thank three anonymous reviewers for their helpful comments.

Conflicts of Interest: The authors declare that they have no conflict of interest.

References

1. Frankham, R.; Ballou, J.D.; Ralls, K.; Eldridge, M.D.B.; Dudash, M.R.; Fenster, C.B.; Lacy, R.C.; Sunnucks, P. *Genetic Management of Fragmented Animal and Plant Populations*; Oxford University Press: Oxford, UK, 2017; p. 432.
2. Fischer, J.; Lindenmayer, D.B. An assessment of the published results of animal relocations. *Biol. Conserv.* **2000**, *96*, 1–11. [CrossRef]
3. Germano, J.M.; Bishop, P.J. Suitability of Amphibians and Reptiles for Translocation. *Conserv. Biol.* **2009**, *23*, 7–15. [CrossRef] [PubMed]
4. Ottewell, K.; Dunlop, J.; Thomas, N.; Morris, K.; Coates, D.; Byrne, M. Evaluating success of translocations in maintaining genetic diversity in a threatened mammal. *Biol. Conserv.* **2014**, *171*, 209–219. [CrossRef]
5. Perez, I.; Anadon, J.D.; Diaz, M.; Nicola, G.G.; Tella, J.L.; Gimenez, A. What is wrong with current translocations? A review and a decision-making proposal. *Front. Ecol. Environ.* **2012**, *10*, 494–501. [CrossRef]
6. Sheean, V.A.; Manning, A.D.; Lindenmayer, D.B. An assessment of scientific approaches towards species relocations in Australia. *Austral Ecol.* **2012**, *37*, 204–215. [CrossRef]
7. Weeks, A.R.; Sgro, C.M.; Young, A.G.; Frankham, R.; Mitchell, N.J.; Miller, K.A.; Byrne, M.; Coates, D.J.; Eldridge, M.D.B.; Sunnucks, P.; et al. Assessing the benefits and risks of translocations in changing environments: A genetic perspective. *Evol. Appl.* **2011**, *4*, 709–725. [CrossRef]
8. Short, J. Predation by feral cats key to the failure of a long-term reintroduction of the western barred bandicoot (*Perameles bougainville*). *Wildl. Res.* **2016**, *43*, 38–50. [CrossRef]
9. Brashares, J.S.; Werner, J.R.; Sinclair, A.R.E. Social 'meltdown' in the demise of an island endemic: Allee effects and the Vancouver Island marmot. *J. Anim. Ecol.* **2010**, *79*, 965–973. [CrossRef] [PubMed]
10. IUCN/SSC. *Guidelines for Reintroductions and Other Conservation Translocations*; Version 1.0; IUCN Special Survival Commission: Gland, Switzerland; p. viiii + 57.

11. Saltz, D. A long-term systematic approach to planning reintroductions: The Persian fallow deer and the Arabian oryx in Israel. *Anim. Conserv.* **1998**, *1*, 245–252. [CrossRef]

12. Holzapfel, S.A.; Robertson, H.A.; McLennan, J.A.; Sporle, W.; Hackwell, K.; Impey, M. Kiwi (*Apteryx* spp.) recovery plan 2008–2018. In *Threatened Species Recovery Plan*; Department of Conservation: Wellington, New Zealand, 2008; Volume 60.

13. Lloyd, B.D.; Powlesland, R.G. The decline of kakapo (*Strigops habroptilus*) and attempts at conservation by translocation. *Biol. Conserv.* **1994**, *69*, 75–85. [CrossRef]

14. Weeks, A.R.; Heinze, D.; Perrin, L.; Stoklosa, J.; Hoffmann, A.A.; van Rooyen, A.; Kelly, T.; Mansergh, I. Genetic rescue increases fitness and aids rapid recovery of an endangered marsupial population. *Nat. Comm.* **2017**, *8*, 1–6. [CrossRef] [PubMed]

15. Frankham, R. Genetics and extinction. *Biol. Conserv.* **2005**, *126*, 131–140. [CrossRef]

16. Allendorf, F.W. Genetic Drift and the Loss of Alleles Versus Heterozygosity. *Zoo Biol.* **1986**, *5*, 181–190. [CrossRef]

17. Gilpin, M.E.; Soule, M.E. Minimum viable populations: Processes of species extinction. In *Conservation Biology: The Science of Scarcity and Diversity*; Gilpin, M.E., Soule, M.E., Eds.; Sinauer: Sunderland, MA, USA, 1986; pp. 19–24.

18. Lacy, R.C. Lessons from 30 years of population viability analysis of wildlife populations. *Zoo Biol.* **2019**, *38*, 67–77. [CrossRef]

19. Tracy, L.N.; Wallis, G.P.; Efford, M.G.; Jamieson, I.G. Preserving genetic diversity in threatened species reintroductions: How many individuals should be released? *Anim. Conserv.* **2011**, *14*, 439–446. [CrossRef]

20. Pacioni, C.; Wayne, A.F.; Page, M. Guidelines for genetic management in mammal translocation programs. *Biol. Conserv.* **2019**, *237*, 105–113. [CrossRef]

21. Grueber, C.E.; Fox, S.; McLennan, E.A.; Gooley, R.M.; Pemberton, D.; Hogg, C.J.; Belov, K. Complex problems need detailed solutions: Harnessing multiple data types to inform genetic management in the wild. *Evol. Appl.* **2019**, *12*, 280–291. [CrossRef]

22. Kelly, E.; Phillips, B. How many and when? Optimising targeted gene flow for a step change in the environment. *Ecol. Lett.* **2019**, *22*, 447–457. [CrossRef]

23. Ramalho, C.E.; Ottewell, K.M.; Chambers, B.K.; Yates, C.J.; Wilson, B.A.; Bencini, R.; Barrett, G. Demographic and genetic viability of a medium-sized ground-dwelling mammal in a fire prone, rapidly urbanizing landscape. *PLoS ONE* **2018**, *13*, e0191190. [CrossRef]

24. Woinarski, J.C.Z.; Burbidge, A.A.; Harrison, P.L. Ongoing unraveling of a continental fauna: Decline and extinction of Australian mammals since European settlement. *Proc. Natl. Acad. Sci. USA* **2015**, *112*, 4531–4540. [CrossRef]

25. Burbidge, A.A.; McKenzie, N.L. Patterns in the Modern Decline of Western-Australia Vertebrate Fauna—Causes and Conservation Implications. *Biol. Conserv.* **1989**, *50*, 143–198. [CrossRef]

26. Cardillo, M.; Bromham, L. Body size and risk of extinction in Australian mammals. *Conserv. Biol.* **2001**, *15*, 1435–1440. [CrossRef]

27. Chisholm, R.; Taylor, R. Null-hypothesis significance testing and the critical weight range for Australian mammals. *Conserv. Biol.* **2007**, *21*, 1641–1645. [CrossRef] [PubMed]

28. Woinarski, J.C.Z. Critical-weight-range marsupials in northern Australia are declining: A commentary on Fisher et al. (2014) 'The current decline of tropical marsupials in Australia: Is history repeating?'. *Glob. Ecol. Biogeogr.* **2015**, *24*, 118–122. [CrossRef]

29. Molloy, S.W.; Davis, R.A.; Van Etten, E.J.B. Species distribution modelling using bioclimatic variables to determine the impacts of a changing climate on the western ringtail possum (*Pseudocheirus occidentals*; Pseudocheiridae). *Environ. Conserv.* **2014**, *41*, 176–186. [CrossRef]

30. Reckless, H.J.; Murray, M.; Crowther, M.S. A review of climatic change as a determinant of the viability of koala populations. *Wildl. Res.* **2017**, *44*, 458–470. [CrossRef]

31. Shortridge, G.C. An account of the geographical distribution of the Marsuprals and Monotremes of South-West Australia, having special reference to the specimens collected during the Balston Expedition of 1904-1907. *Proc. Zool. Soc. Lond.* **1909**, *1909*, 803–848.

32. Thomas, N.; (Department of Biodiversity, Conservation and Attractions, Perth, WA, Australia). Personal communication, 2018.

33. Chapman, T.F.; Sims, C.; Thomas, N.D.; Reinhold, L. Assessment of mammal populations on Bernier and Dorre Island 2006–2013. In *Report for the Department of Parks and Wildlife*; Department of Parks and Wildlife: Kensington, Australia, 2015; p. 56.

34. Short, J.; Turner, B.; Majors, C.; Leone, J. The fluctuating abundance of endangered mammals on Bernier and Dorre Islands, Western Australia—conservation implications. *Aust. Mammal.* **1997**, *20*, 53–61.

35. Hardman, B.; Moro, D.; Calver, M. Direct evidence implicates feral cat predation as the primary cause of failure of a mammal reintroduction programme. *Ecol. Manag. Restor.* **2016**, *17*, 152–158. [CrossRef]

36. Morris, K.; Sims, C.; Himbeck, K.; Christensen, P.; Sercombe, N.; Ward, B.; Noakes, N. Project Eden—fauna recovery on Peron Peninsula, Shark Bay: Western Shield review—February 2003. *Conserv. Sci. West. Aust.* **2004**, *5*, 202–234.

37. Short, J.; Turner, B. The Distribution and Abundance of the Banded and Rufous Hare-Wallabies, *Lagostrophus fasciatus* and *Lagorchestes hirsutus*. *Biol. Conserv.* **1992**, *60*, 157–166. [CrossRef]

38. Ruykys, L.; Smith, M.; Kanowski, J. *Translocation of Banded Hare-wallabies (Lagostrophus fasciatus) to Mt Gibson Sanctuary and Faure Island, WA*; Department of Biodiversity, Conservation and Attractions: Kensington, Australia, 2017; p. 44.

39. Short, J.; Hide, A. Successful reintroduction of the brushtail possum to Wadderin Sanctuary in the eastern wheatbelt of Western Australia. *Aust. Mammal.* **2014**, *36*, 229–241. [CrossRef]

40. Short, J.; (Wildlife Research and Management Pty Ltd., Perth, WA, Australia). Personal communication, 2020.

41. Morris, K.; Page, M.; Thomas, N.; Ottewell, K. *A strategic framework for the reconstruction and conservation of the vertebrate fauna of Dirk Hartog Island 2016–2030*; Department of Parks and Wildlife: Perth, Australia, 2017; p. 26.

42. Harris, R.M.B.; Beaumont, L.J.; Vance, T.R.; Tozer, C.R.; Remenyi, T.A.; Perkins-Kirkpatrick, S.E.; Mitchell, P.J.; Nicotra, A.B.; McGregor, S.; Andrew, N.R.; et al. Biological responses to the press and pulse of climate trends and extreme events. *Nat. Clim. Chang.* **2018**, *8*, 579–587. [CrossRef]

43. Sunnucks, P.; Hales, D.F. Numerous transposed sequences of mitochondrial cytochrome oxidase I-II in aphids of the genus Sitobion (Hemiptera: Aphididae). *Mol. Biol. Evol.* **1996**, *13*, 510–524. [CrossRef]

44. Taylor, A.C.; Cooper, D.W. A set of tammar wallaby (*Macropus eugenii*) microsatellites tested for genetic linkage. *Mol. Ecol.* **1998**, *7*, 925–926.

45. Pope, L.C.; Sharp, A.; Moritz, C. Population structure of the yellow-footed rock-wallaby *Petrogale xanthopus* (Gray, 1854) inferred from mtDNA sequences and microsatellite loci. *Mol. Ecol.* **1996**, *5*, 629–640. [CrossRef]

46. Spencer, P.B.S.; Odorico, D.M.; Jones, S.J.; Marsh, H.D.; Miller, D.J. Highly Variable Microsatellites in Isolated Colonies of the Rock-Wallaby (*Petrogale assimilis*). *Mol. Ecol.* **1995**, *4*, 523–525. [CrossRef]

47. Rousset, F. GENEPOP ' 007: A complete re-implementation of the GENEPOP software for Windows and Linux. *Mol. Ecol. Resour.* **2008**, *8*, 103–106. [CrossRef]

48. Guo, S.W.; Thompson, E.A. Performing the exact test of hardy-weinberg proportion for multiple alleles. *Biometrics* **1992**, *48*, 361–372. [CrossRef]

49. Van Oosterhout, C.; Hutchinson, W.F.; Wills, D.P.M.; Shipley, P. MICRO-CHECKER: Software for identifying and correcting genotyping errors in microsatellite data. *Mol. Ecol. Notes* **2004**, *4*, 535–538. [CrossRef]

50. Pritchard, J.K.; Stephens, M.; Donnelly, P. Inference of population structure using multilocus genotype data. *Genetics* **2000**, *155*, 945–959. [PubMed]

51. Evanno, G.; Regnaut, S.; Goudet, J. Detecting the number of clusters of individuals using the software STRUCTURE: A simulation study. *Mol. Ecol.* **2005**, *14*, 2611–2620. [CrossRef] [PubMed]

52. Kopelman, N.M.; Mayzel, J.; Jakobsson, M.; Rosenberg, N.A.; Mayrose, I. Clumpak: A program for identifying clustering modes and packaging population structure inferences across K. *Mol. Ecol. Res.* **2015**, *15*, 1179–1191. [CrossRef] [PubMed]

53. Rosenberg, N.A. DISTRUCT: A program for the graphical display of population structure. *Mol. Ecol. Notes* **2004**, *4*, 137–138. [CrossRef]

54. Peakall, R.; Smouse, P. GenAlEx 6.5: Genetic analysis in Excel. Population genetic software for teaching and research—An update. *Bioinformatics* **2012**, *28*, 2537–2539. [CrossRef] [PubMed]

55. Peakall, R.; Smouse, P.E. GENALEX 6: Genetic analysis in Excel. Population genetic software for teaching and research. *Mol. Ecol. Notes* **2006**, *6*, 288–295. [CrossRef]

56. Kalinowski, S.T. HP-RARE 1.0: A computer program for performing rarefaction on measures of allelic richness. *Mol. Ecol. Notes* **2005**, *5*, 187–189. [CrossRef]

57. Hill, W.G. Estimation of Effective Population-Size from Data on Linkage Disequilibrium. *Genet. Res.* **1981**, *38*, 209–216. [CrossRef]

58. Waples, R.S. A bias correction for estimates of effective population size based on linkage disequilibrium at unlinked gene loci. *Conserv. Genet.* **2006**, *7*, 167–184. [CrossRef]

59. Waples, R.S.; Do, C. Linkage disequilibrium estimates of contemporary Ne using highly variable genetic markers: A largely untapped resource for applied conservation and evolution. *Evol. Appl.* **2010**, *3*, 244–262. [CrossRef]

60. Do, C.; Waples, R.S.; Peel, D.; Macbeth, G.M.; Tillett, B.J.; Ovenden, J.R. NEESTIMATOR v2: Re-implementation of software for the estimation of contemporary effective population size (Ne) from genetic data. *Mol. Ecol. Res.* **2014**, *14*, 209–214. [CrossRef]

61. Piry, S.; Luikart, G.; Cornuet, J.M. BOTTLENECK: A computer program for detecting recent reductions in the effective population size using allele frequency data. *J. Hered.* **1999**, *90*, 502–503. [CrossRef]

62. Cornuet, J.M.; Luikart, G. Description and power analysis of two tests for detecting recent population bottlenecks from allele frequency data. *Genetics* **1996**, *144*, 2001–2014.

63. Luikart, G.; Sherwin, W.B.; Steele, B.M.; Allendorf, F.W. Usefulness of molecular markers for detecting population bottlenecks via monitoring genetic change. *Mol. Ecol.* **1998**, *7*, 963–974. [CrossRef]

64. Luikart, G.; Allendorf, F.W.; Cornuet, J.M.; Sherwin, W.B. Distortion of allele frequency distributions provides a test for recent population bottlenecks. *J. Hered.* **1998**, *89*, 238–247. [CrossRef]

65. Queller, D.C.; Goodnight, K.F. Estimating Relatedness Using Genetic-Markers. *Evolution* **1989**, *43*, 258–275. [CrossRef]

66. Weiser, E.L.; Grueber, C.E.; Jamieson, I.G. AlleleRetain: A program to assess management options for conserving allelic diversity in small, isolated populations. *Mol. Ecol. Res.* **2012**, *12*, 1161–1167. [CrossRef]

67. Weiser, E.L.; Grueber, C.E.; Jamieson, I.G. Simulating Retention of Rare Alleles in Small Populations to Assess Management Options for Species with Different Life Histories. *Conserv. Biol.* **2013**, *27*, 335–344. [CrossRef]

68. R Core Team. *R: A Language and Environment for Statistical Computing*; R Foundation for Statistical Computing: Vienna, Austria, 2019; Available online: https://www.R-project.org/ (accessed on 11 May 2019).

69. Sims, C.; (Department of Biodiversity, Conservation and Attractions, Perth, WA, Australia). Personal communication, 2018.

70. Richards, J.D.; Short, J.; Prince, R.I.T.; Friend, J.A.; Courtenay, J.M. The biology of banded (Lagostrophus fasciatus) and rufous (Lagorchestes hirsutus) hare-wallabies (Diprotodontia: Macropodidae) on Dorre and Bernier Islands, Western Australia. *Wildl. Res.* **2001**, *28*, 311–322. [CrossRef]

71. Lacy, R.C. VORTEX—A computer-simulation model for population viability analysis. *Wildl. Res.* **1993**, *20*, 45–65. [CrossRef]

72. Lacy, R.C.; Pollak, J.P. *Vortex: A Stochastic Simulation of the Extinction Process*; Version 10.2.9; Chicago Zoological Society: Brookfield, IL, USA, 1993; Available online: http://www.vortex10.org/Vortex10.aspx (accessed on 11 May 2019).

73. Eldridge, M.D.B.; King, J.M.; Loupis, A.K.; Spencer, P.B.S.; Taylor, A.C.; Pope, L.C.; Hall, G.P. Unprecedented low levels of genetic variation and inbreeding depression in an island population of the black-footed rock-wallaby. *Conserv. Biol.* **1999**, *13*, 531–541. [CrossRef]

74. Chapuis, M.P.; Estoup, A. Microsatellite null alleles and estimation of population differentiation. *Mol. Biol. Evol.* **2007**, *24*, 621–631. [CrossRef] [PubMed]

75. Frankham, R. Do island populations have less genetic variation than mainland populations? *Heredity* **1997**, *78*, 311–327. [CrossRef] [PubMed]

76. Wright, S. Evolution in Mendelian populations. *Genetics* **1931**, *16*, 0097–0159.

77. Pacioni, C.; Wayne, A.F.; Spencer, P.B.S. Genetic outcomes from the translocations of the critically endangered woylie. *Curr. Zool.* **2013**, *59*, 294–310. [CrossRef]

78. Thavornkanlapachai, R.; Mills, H.R.; Ottewell, K.; Dunlop, J.; Sims, C.; Morris, K.; Donaldson, F.; Kennington, W.J. Mixing Genetically and Morphologically Distinct Populations in Translocations: Asymmetrical Introgression in A Newly Established Population of the Boodie (*Bettongia lesueur*). *Genes* **2019**, *10*, 729. [CrossRef]

79. Smith, S.; Hughes, J. Microsatellite and mitochondrial DNA variation defines island genetic reservoirs for reintroductions of an endangered Australian marsupial, *Perameles bougainville*. *Conserv. Genet.* **2008**, *9*, 547–557. [CrossRef]

80. Eldridge, M.D.B.; Neaves, L.E.; Spencer, P.B.S. Genetic analysis of three remnant populations of the rufous hare-wallaby (*Lagorchestes hirsutus*) in arid Australia. *Aust. Mammal.* **2019**, *41*, 123–131. [CrossRef]

81. Frankham, R. Effective Population Size Adult Population Size Ratios in Wildlife—A Review. *Genet. Res.* **1995**, *66*, 95–107. [CrossRef]

82. Nunney, L. The influence of variation in female fecundity on effective population size. *Biol. J. Linn. Soc.* **1996**, *59*, 411–425. [CrossRef]

83. Waples, R.S.; Luikart, G.; Faulkner, J.R.; Tallmon, D.A. Simple life-history traits explain key effective population size ratios across diverse taxa. *Proc. R. Soc. B Biol. Sci.* **2013**, *280*. [CrossRef] [PubMed]

84. Churchill, D.M. Late Quaternary eustatic changes in the Swan River district. *J. R. Soc. West. Aust.* **1959**, *42*, 53–55.

85. Short, J.; Turner, B.; Majors, C. *The Distribution, Relative Abundance, and Habitat Preferences of Rare Macropods and Bandicoots on Barrow, Boodie, Bernier and Dorre Islands*; CSIRO Division of Wildlife and Ecology: Midland, Australia, 1989; p. 64.

86. Goldstein, D.B.; Pollock, D.D. Launching microsatellites: A review of mutation processes and methods of phylogenetic inference. *J. Hered.* **1997**, *88*, 335–342. [CrossRef] [PubMed]

87. Smith, M.; (Australian Wildlife Conservancy, Perth, WA, Australia). Personal communication, 2020.

88. Sonsthagen, S.A.; Wilson, R.E.; Underwood, J.G. Genetic implications of bottleneck effects of differing severities on genetic diversity in naturally recovering populations: An example from Hawaiian coot and Hawaiian gallinule. *Ecol. Evol.* **2017**, *7*, 9925–9934. [CrossRef] [PubMed]

89. Heber, S.; Briskie, J.V. Population Bottlenecks and Increased Hatching Failure in Endangered Birds. *Conserv. Biol.* **2010**, *24*, 1674–1678. [CrossRef]

90. Easton, L.J.; Bishop, P.J.; Whigham, P.A. Balancing act: Modelling sustainable release numbers for translocations. *Anim. Conserv.* **2019**. [CrossRef]

91. Allendorf, F.W.; Ryman, N. The role of genetics in population viability analysis. In *Population Viability Analysis*; Beissinger, S.R., McCullough, D.R., Eds.; University of Chicago Press: Chicago, IL, USA, 2002; pp. 50–85.

92. Pacioni, C.; Rafferty, C.; Morley, K.; Stevenson, S.; Chapman, A.; Wickins, M.; Verney, T.; Deegan, G.; Trocini, S.; Spencer, P.B.S. Augmenting the conservation value of rehabilitated wildlife by integrating genetics and population modeling in the post-rehabilitation decision process. *Curr. Zool.* **2018**, *64*, 593–601. [CrossRef]

93. Allendorf, F.W.; Leary, R.F.; Spruell, P.; Wenburg, J.K. The problems with hybrids: Setting conservation guidelines. *Trends Ecol. Evol.* **2001**, *16*, 613–622. [CrossRef]

94. Edmands, S. Between a rock and a hard place: Evaluating the relative risks of inbreeding and outbreeding for conservation and management. *Mol. Ecol.* **2007**, *16*, 463–475. [CrossRef]

95. Frankham, R.; Ballou, J.D.; Eldridge, M.D.B.; Lacy, R.C.; Ralls, K.; Dudash, M.R.; Fenster, C.B. Predicting the Probability of Outbreeding Depression. *Conserv. Biol.* **2011**, *25*, 465–475. [CrossRef]

96. Armbruster, P.; Bradshaw, W.E.; Steiner, A.L.; Holzapfel, C.M. Evolutionary responses to environmental stress by the pitcher-plant mosquito, *Wyeomyia smithii*. *Heredity* **1999**, *83*, 509–519. [CrossRef] [PubMed]

97. Edmands, S. Heterosis and outbreeding depression in interpopulation crosses spanning a wide range of divergence. *Evolution* **1999**, *53*, 1757–1768. [CrossRef] [PubMed]

98. Marr, A.B.; Keller, L.F.; Arcese, P. Heterosis and outbreeding depression in descendants of natural immigrants to an inbred population of song sparrows (*Melospiza melodia*). *Evolution* **2002**, *56*, 131–142. [CrossRef] [PubMed]

99. Tymchuk, W.E.; Sundstrom, L.F.; Devlin, R.H. Growth and survival trade-offs and outbreeding depression in rainbow trout (*Oncorhynchus mykiss*). *Evolution* **2007**, *61*, 1225–1237. [CrossRef] [PubMed]

Publisher's Note: MDPI stays neutral with regard to jurisdictional claims in published maps and institutional affiliations.

Article

Applying Population Viability Analysis to Inform Genetic Rescue That Preserves Locally Unique Genetic Variation in a Critically Endangered Mammal

Joseph P. Zilko [1,*], Dan Harley [2], Alexandra Pavlova [1] and Paul Sunnucks [1]

[1] School of Biological Sciences, Monash University, Clayton, VIC 3800, Australia; alexandra.pavlova@monash.edu (A.P.); paul.sunnucks@monash.edu (P.S.)

[2] Zoos Victoria, Parkville, VIC 3052, Australia; dharley@zoo.org.au

* Correspondence: joseph.zilko@monash.edu

Abstract: Genetic rescue can reduce the extinction risk of inbred populations, but it has the poorly understood risk of 'genetic swamping'—the replacement of the distinctive variation of the target population. We applied population viability analysis (PVA) to identify translocation rates into the inbred lowland population of Leadbeater's possum from an outbred highland population that would alleviate inbreeding depression and rapidly reach a target population size (N) while maximising the retention of locally unique neutral genetic variation. Using genomic kinship coefficients to model inbreeding in Vortex, we simulated genetic rescue scenarios that included gene pool mixing with genetically diverse highland possums and increased the N from 35 to 110 within ten years. The PVA predicted that the last remaining population of lowland Leadbeater's possum will be extinct within 23 years without genetic rescue, and that the carrying capacity at its current range is insufficient to enable recovery, even with genetic rescue. Supplementation rates that rapidly increased population size resulted in higher retention (as opposed to complete loss) of local alleles through alleviation of genetic drift but reduced the frequency of locally unique alleles. Ongoing gene flow and a higher N will facilitate natural selection. Accordingly, we recommend founding a new population of lowland possums in a high-quality habitat, where population growth and natural gene exchange with highland populations are possible. We also recommend ensuring gene flow into the population through natural dispersal and/or frequent translocations of highland individuals. Genetic rescue should be implemented within an adaptive management framework, with post-translocation monitoring data incorporated into the models to make updated predictions.

Keywords: genetic swamping; genetic rescue; locally unique alleles; translocation; population viability analysis; population genetics; marsupial; adaptive potential; Leadbeater's possum; *Gymnobelideus leadbeateri*

Citation: Zilko, J.P.; Harley, D.; Pavlova, A.; Sunnucks, P. Applying Population Viability Analysis to Inform Genetic Rescue That Preserves Locally Unique Genetic Variation in a Critically Endangered Mammal. *Diversity* **2021**, *13*, 382. https://doi.org/10.3390/d13080382

Academic Editors: Kym Ottewell and Margaret Byrne

Received: 30 June 2021
Accepted: 12 August 2021
Published: 16 August 2021

Publisher's Note: MDPI stays neutral with regard to jurisdictional claims in published maps and institutional affiliations.

1. Introduction

Many threatened species are restricted to small and isolated populations, which places them at elevated risk of inbred matings and loss of beneficial genetic variation by genetic drift [1,2]. Small and isolated populations of diploid species that usually outbreed nearly always exhibit reduced fitness as a consequence of inbreeding (i.e., inbreeding depression) [3]. Usually concurrent with inbreeding depression, the loss of beneficial variation and the accumulation of harmful variation further reduce the population mean fitness [4]. Together with the loss of genetic variation that would otherwise facilitate adaptation to future environmental conditions, these issues reduce population viability [1,5]. If inbred populations remain isolated, extinction becomes increasingly likely due to the 'extinction vortex' effect, wherein a shrinking population size is reinforced by increasing inbreeding depression and fitness loss [6,7].

Translocations of individuals from other populations can increase genetic diversity of isolated populations and elevate population fitness by reducing inbreeding depression and alleviating the harmful effects of beneficial variation being lost from the population (i.e., causing genetic rescue), thus increasing the persistence and adaptive capacity of populations [2,8,9]. Such genetic management is a highly effective conservation tool; when appropriate sources of genetic augmentation are used, outcrossing increases fitness in the vast majority of inbred populations that usually outbreed, and an enhanced ability to adapt can greatly increase population persistence in changing environments [10].

Despite its demonstrated effectiveness, genetic rescue has not been widely implemented, partly owing to concerns about 'genetic swamping' and a consequent loss of local adaptation and distinctiveness [11–13]. Genetic swamping can be defined as the replacement of a population's locally adaptive genetic variation with that of immigrants [13]. This definition may be extended to include apparently neutral alleles within populations because distinguishing adaptive from non-adaptive variation can be extremely challenging, and such variation can contribute to future adaptive potential [14]. An extreme example of genetic swamping occurs in some hybridising species, in which hybrid vigour (heterosis) causes the extirpation of the original locally adapted population and their alleles [15]. A subtler form of genetic swamping occurs when locally unique genetic variation is replaced during the immigration of conspecifics [16]. Concerns about genetic swamping tend to promote conservation strategies that maintain genetic distinctiveness of isolated populations rather than evaluating alternative management options in a risk-benefit framework, but a different strategy might have a better outcome for the stated conservation goals [17]. On one hand, species-wide genetic variation could be lost if translocation strategies cause the replacement of locally unique alleles, potentially reducing the adaptive potential and increasing extinction risk [13,18–20]. Conversely, the genetic distinctiveness of small and isolated populations is often an artefact of genetic stochasticity rather than local adaptation, and if 'uniqueness' reflects harmful variation that would have been removed by selection in larger and/or more connected populations, maintaining the distinctiveness of small populations can even exacerbate extinction risk [21].

Effective population sizes (N_e) >100 are needed to limit the mean inbreeding coefficient (the probability that any two alleles are identical by descent) to $F < 0.1$ over five generations [10]. Smaller and more inbred populations need high supplementation rates to reverse inbreeding, which could elevate the risk of swamping [1]. Nevertheless, alleviating inbreeding depression while avoiding the replacement of the local gene pool may be achievable by appropriately managing gene flow during genetic rescue. Even under considerable gene flow during intensive genetic rescue, selection can maintain locally adaptive genetic variation in the face of potential swamping [22]. However, selection is generally weaker in small populations because genetic drift has a greater influence on allele frequencies [23]. Preserving genome-wide variation, including variation that is currently neutral, also enhances future adaptive potential [14,24,25]. Thus, it is of interest to estimate how much locally unique genome-wide variation will persist in a population under genetic rescue, especially if candidate populations for genetic rescue are small. Accordingly, poor understanding of swamping during genetic rescue is a significant knowledge gap [13].

Population demographic and genetic models that account for inbreeding depression and the genetic contribution of outbred immigrants could be useful for designing genetic rescue regimes that help retain locally unique alleles. Forward-based computer simulations can project the changes in the genetic profile of populations under different demographic and population genetic scenarios [26]. Such simulations have been used to estimate the impacts of inbreeding depression and predict changes in the viability of inbred populations after genetic rescue [14,27]. PVAs can be used to predict how much of the genetic variation from a source population can be retained when founding a new population or supplementing an existing population [28–30]. However, the preservation of alleles unique to a local population that is supplemented with unrelated individuals has been less scrutinised, and fitness disparities between outbred immigrants and inbred locals can cause the replacement of local variation [31]. Incorporating inbreeding and inbreeding depression into models accounts for the reduced contribution to the gene pool of inbred locals relative to immigrants. Models that consider locally unique variation and the lower fitness of inbred locals relative to immigrants could therefore be useful for designing genetic rescue programs that improve population viability without replacing the local gene pool.

Leadbeater's possum (*Gymnobelideus leadbeateri*) is a critically endangered Australian marsupial and a strong candidate for genetic rescue. Genetic rescue of this species could be informed by population viability analysis (PVA) modelling the effect of various translocation scenarios on the retention of unique alleles when the recipient gene pool has reduced fitness due to severe inbreeding. The species has suffered extensive loss and deterioration of habitat caused by vegetation clearing, bushfires, and logging, and has a highly restricted distribution (Figure 1) within a 70 × 95 km area within the Central Highlands of Victoria, Australia, with a neighbouring lowland isolate [32,33]. The single population of Leadbeater's possum outside of the Central Highlands is localised entirely within a narrow 6 km stretch of lowland swamp forest habitat in the Yellingbo Nature Conservation Reserve [34]. The Yellingbo population is genetically distinct from highland populations and is thought to represent the sole remnant of a previously widespread evolutionary lineage adapted to lowland swamp [32,35,36]. The geographic isolation of the Yellingbo population makes contemporary gene flow with other populations highly unlikely [36]. This population had greatly reduced genetic diversity compared to all conspecific populations and exhibited substantial inbreeding depression for survival to sexual maturity, longevity, probability of breeding during lifetime, and lifetime reproductive output [37]. Inbreeding depression likely contributed to the decline from ~120 individuals in 1997 to 35 individuals in 2017. Heterozygosity concurrently declined by 12%, putatively limiting the capacity of the population to adapt to environmental change and prevent the expression of deleterious variation [37]. Furthermore, suitable habitat for Leadbeater's possum has declined in extent and quality, mainly because of habitat succession and altered hydrology [32]. Based on monotonic decline in population size, presence of inbreeding depression, and deteriorating habitat condition, the extinction of the Yellingbo population is highly likely without intervention [37,38]. Translocation of lowland Yellingbo individuals to high-quality habitat beyond Yellingbo and outcrossing with genetically diverse highland animals is planned to prevent the extinction of this population. Sourcing migrants for genetic rescue of the lowland population from highland populations is the only available option, which has been assessed to have a low risk of outbreeding depression [37]. Although genetic variation unique to the lowland population at Yellingbo has been characterised with microsatellite markers and short mitochondrial sequences [36], it was not assessed with genome-wide markers, which should better approximate functional and standing variation [24,25]. Doing so would be an important step in planning translocation strategies that do not replace the local gene pool.

We aimed to design a translocation regime of lowland and highland Leadbeater's possums into new habitat that would rapidly increase population size, demographic stability, and genetic diversity, while providing for the retention of unique lowland alleles. We predicted the level of retention of alleles unique to the lowland population, incorporating inbreeding depression to reflect the potential fitness advantage of translocated outbred individuals and their descendants over inbred locals. We used genome-wide SNP markers to detect genomic variation unique to the lowland population, which was then monitored across simulated scenarios. We also used genome-wide SNPs to model inbreeding and genetic diversity within simulated populations. The following objectives were targeted:

(1) Predict the trajectory of the lowland population at Yellingbo without genetic rescue;
(2) Determine whether genetic rescue can retain genetic variation unique to the lowland population;
(3) Assess whether more rapid genetic rescue presents a greater risk of uniquely lowland variation being replaced.

Our results will inform genetic rescue of the last extant population of the lowland lineage of this critically endangered species. More generally, our approach for designing genetic rescue regimes to retain locally unique variation should be useful for other conservation programs.

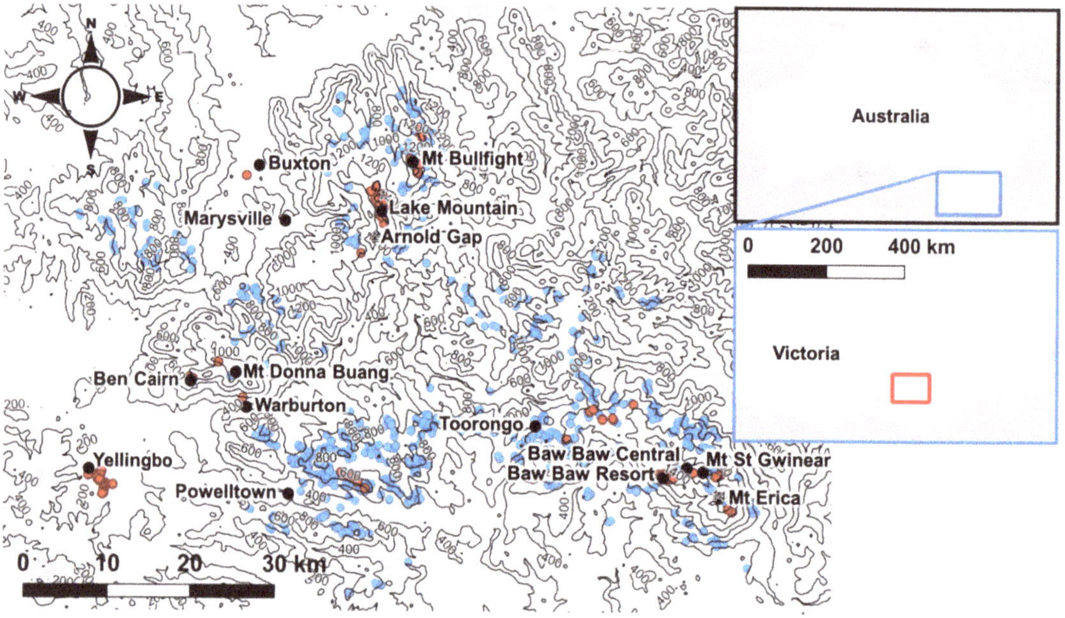

Figure 1. Map of sampling locations for genetic samples used in this study (red dots). The blue dots are observations of the species from 2010–2020, indicating the contemporary range. The extent of the sampling area is indicated by the red rectangle on the inset. Contour lines indicate elevation above sea level in metres.

2. Materials and Methods

2.1. Study System and Sampling

The lowland population at Yellingbo was extensively surveyed from 1997–2019, with mark–recapture data available for >70% of the population in any given year [37,39]. These data provided estimates of key demographic parameters, such as mortality and reproduction rates (see below). Genetic samples (n = 304) were collected from Yellingbo during surveys from 1997 to 2001 and 2011 to 2019 [36,37]. A further 216 individuals were genetically sampled throughout the species' range in the Central Highlands (Figure 1), against which to estimate the locally unique variation at Yellingbo. These included 113 individuals from Lake Mountain—the thoroughly sampled population used to represent allele frequencies of a donor highland population in simulations. For most samples, a small (~2 mm²) tissue biopsy of skin and cartilage was taken from the outer pinna of the ear and stored in ethanol prior to extraction. Skin samples were also taken from animals found dead or from the pelts of preserved museum specimens.

2.2. Defining Conservation Targets

Conservation programs guided by PVAs require clearly defined criteria of success [40]. We defined a strategy as successful if it resulted in a population of at least 110 individuals. This target was chosen to recreate apparent demographic stability observed at Yellingbo in the 1990s before the rapid deterioration of habitat conditions and concurrent population decline. The limited availability of lowland habitat also means that founded populations must necessarily be <120 individuals in size, with translocations maintaining population size and genetic diversity. This target represents a first (and urgent) milestone to prevent the extinction of the lowland population; however, further population expansion would be required thereafter. We assumed that genetic rescue would begin with 20 Yellingbo founders, which represents 57% of the population in 2019.

2.3. General Assumptions of the PVA Model

All PVAs were performed using the software package Vortex v.10.3.5.0 [41], an individual-based Monte Carlo simulation program that models the impact of deterministic and stochastic forces on animal populations. Demographic, environmental, and genetic processes were incorporated by inputting parameter estimates for mortality and fecundity, along with environmental variability. For a range of scenarios (Table 1), the program tracked individuals from birth to death, with reproduction and mortality occurring at pre-defined probabilities, derived from actual population data (below). Environmental catastrophes were not included in the model. Due to the uncertainty of environmental effects on fitness, environmental effects were minimised by setting the standard deviation for mortality and reproduction parameters to 10% of their mean value (Table 2). This was possible only for the parameters modelled using mean and standard deviation values.

Table 1. Scenarios simulated in Vortex. Initial lowland population size (N), carrying capacity (K), whether or not the starting population was made inbred by providing a kinship matrix (Inbred), whether inbreeding depression was modelled, and whether estimated or adjusted values for juvenile (0–1 y) mortality were used is indicated. Supplementation rate is the number of individuals supplemented (by translocation) into the population per year. Supplementation duration is the number of years the population was supplemented. Simulation duration is the number of years simulations were run. Source is the population that animals were sourced from for supplementation.

Scenario	N	K	Inbred	Inbreeding Depression	Juvenile Mortality	Supplementation Rate	Supplementation Duration (Years)	Simulation Duration (Years)	Source
1990s Condition	110	120	No	Yes and no	Estimated	0	NA	50	NA
		1000	No	Yes and no	Estimated	0	NA	50	NA
2019 Trajectory	35	40	Yes	Yes	6%	0	NA	50	NA
		120	Yes	Yes	6%	0	NA	50	NA
Demographic Rescue	20	1000	Yes	Yes	6%	2; 4; 6; 8; 10	50	50	Lowland
Genetic Rescue	20	1000	Yes	Yes	6%	2; 4; 6; 8; 10	50	50	Highland
Genetic Swamping Test	20	1000	Yes	Yes	6%	2; 4; 6; 8; 10	50	10; 20; 30; 40; 50	Highland

Table 2. Biological parameters used in Vortex simulations. The study that provided each parameter estimate is given.

Parameter	Value (SD)	Source
Diploid lethal equivalents	6.29	[42]
Percent inbreeding due to lethal recessives	50	[43]
Environmental variation correlation between reproduction and survival	0	Not used
Environmental variation correlation among populations	0	Not used
Age producing first offspring (years)	2	[39]
Maximum lifespan (years)	9	[37]
Maximum litters per year	2 *	[39]
Maximum progeny per litter	2	[39]
Maximum age of reproduction (years)	9	[37]
Sex ratio at birth	01:01	[39]
Percent adult females breeding	66.6 (6.7)	[39]
Distribution of litters per year		
0 litter	0	[44]

Table 2. *Cont.*

Parameter	Value (SD)	Source
1 litter	7	[44]
2 litter	93	[44]
Distribution of offspring per litter		
1 offspring	45	[39]
2 offspring	55	[39]
Mortality		
0–1 years (males)	42 (4.2) †	[44]
0–1 years (females)	41 (4.1) †	[44]
1–2 years (males)	17 (1.7)	[44]
1–2 years (females)	22 (2.2)	[44]
Annual >2 years (males)	29 (2.9)	[37]
Annual >2 years (females)	29 (2.9)	[37]
Percent males in breeding pool	100	

* Up to three litters per year can be produced; however, this only occurs if a previous litter is lost. † 6% was added to early mortality estimates because these did not include partial litter loss, except for *1990s Condition* and *2019 Trajectory* (main text).

Genetic data were incorporated into all but one analysis (*1990s Condition* being the exception) as the input allele frequencies and pairwise kinships among individuals (detailed methods below). Alleles were assumed to be neutral and randomly assorting. Vortex uses pairwise kinships to calculate inbreeding coefficients for individuals and their offspring. Inbreeding depression was included in some models as a reduced survival probability in the first year of life (the default in Vortex). Inbreeding depression is caused by an increased expression of deleterious recessive alleles and a loss of heterozygote advantage with increasing autozygosity [45]. Lethal equivalents represent recessive alleles that compromise fitness when homozygous in an individual, with a single homozygous lethal equivalent imposing a fitness cost equivalent to the death of an individual [3]. Decreases in individual survival probability are determined by the number of lethal equivalents, the relative contribution of lethal recessive alleles and the heterozygote advantage to inbreeding depression, and the individual inbreeding coefficient. The default value of 6.29 diploid lethal equivalents for first year survival was used. This value was used in preference to estimates of lethal equivalents in the lowland Leadbeater's possum population obtained in a previous study because these estimates were found to have wide confidence intervals and varied between different inbreeding coefficients [37]. Vortex randomly assigns lethal equivalents to each founder at the beginning of simulations. In Vortex, homozygosity for any lethal recessive always results in mortality in the first year of life. The relative contribution of lethal and sub-lethal recessive alleles to inbreeding depression is not well understood, and therefore, we specified 50% inbreeding depression due to lethal recessive alleles, a classic empirical estimate from fruit flies [43]. The reduction in first year survival caused by reduced heterozygote advantage is calculated by applying an exponential equation that includes the inbreeding coefficient of an individual.

2.4. Genotyping

DNA was extracted by salting-out [46] or using the DNeasy Blood and Tissue Kit following Qiagen's (the manufacturer) instructions. The preparation of reduced-representation genome libraries, sequencing, and the discovery of genome-wide single nucleotide polymorphism loci (SNPs) was carried out using the DArTseq complexity reduction methodology of Diversity Arrays Technology [47] and described in a previous study [37]. This generated 13368 SNP loci that were then filtered for downstream analysis using the *R* package *dartR* [48], as explained below. The function *gl.sexlinkage* was used to detect putatively sex-linked loci, which were removed from the dataset. Loci containing more than one SNP were thinned to retain the SNP with the highest polymorphism information. Locus reproducibility was determined during marker discovery from technical replicates representing 25% of all data, and loci with average reproducibility <100% were removed from the dataset to minimise error rates.

Three genetic datasets were then created by sub-setting the data by sampling location and applying different quality-filtering criteria in *dartR* to meet the requirements of each downstream analysis. The first dataset of 4218 loci, created by retaining individuals from all sampling locations and removing loci with a call rate of <85%, was used to detect alleles unique to the lowland population at Yellingbo. The second dataset of 1066 loci, created by retaining only the Lake Mountain and Yellingbo populations and removing loci with a call rate of <100%, was used during simulations to represent genetic variation at Yellingbo and a thoroughly sampled representative highland population. The third dataset of 1076 loci, created by retaining only Yellingbo individuals and removing loci with a call rate of <99%, was used to calculate pairwise kinships at Yellingbo.

2.5. Detecting Alleles Unique to the Lowland Population

Alleles occurring only within the lowland population at Yellingbo (locally unique alleles) were detected by comparing the allele frequencies of all highland individuals pooled together with allele frequencies of lowland individuals sampled at Yellingbo from 1997–2019. The package *poppr* [49] was used to report unique alleles and counts of each locally unique allele in the population. Loci with locally unique alleles found as a single copy in only one Yellingbo individual were removed, as these could represent genotyping errors or rare deleterious mutations. Allele frequencies calculated for loci with locally unique alleles for Yellingbo individuals sampled in 2011–2019 were used in genetic rescue simulations.

2.6. Pairwise Kinship Estimation

Inbreeding among individuals in the population was estimated using pairwise kinships, which measure the proportion of the genome that is identical by shared ancestry between pairs of individuals [50]. Kinships calculated using genetic markers often better represent genomic similarity than do pedigree-derived estimates [51]. Pairwise kinships were calculated with the *beta.dosage* function in the *R* package *hierfstat* [52] using 1076 loci and all Yellingbo genotypes from 1997 to 2019. This kinship estimator has a mean of zero for the focal population, with negative values representing below-average kinship and positive values representing above-average kinship [53]. Vortex requires positive kinship values that represent the probability that any two homologous alleles drawn from a pair of individuals are identical by descent [41]. Therefore, we used the transformation described previously [54] to convert calculated kinships to proportions, with the resulting pairwise kinship value for the least-related pair equal to zero. Kinship values for the 2019 population were incorporated into simulations as a matrix. By default, Vortex regards non-genotyped individuals as completely outbred and unrelated, which is unrealistic. Accordingly, five non-genotyped individuals were represented in the kinship matrix with pairwise kinship values set to the mean calculated for the 2019 individuals.

2.7. Survival and Reproduction Parameters Used in Vortex Models

The Vortex models were populated with realistic parameter values (Table 2) using published estimates of reproduction and mortality at different life stages [39]. The age classes for which that study calculated mortality differed from those required by Vortex. As such, mortality from 0 to 1 years was calculated as the sum of total litter loss (9%) for both sexes, plus mortality for each sex from when juveniles first emerged from the pouch to 12 months old. The mean annual adult (>2 years old) mortality rate was estimated by averaging the annual estimates from 1996 to 1999.

Based on the observation of a relatively stable Yellingbo population at ~110 individuals in the 1990s, simulations using realistic demographic parameters were expected to yield a demographically stable population—that is, one that did not change in size by more than 10% over 50 years in the absence of inbreeding depression or environmental catastrophes (see scenario *1990s Condition* below). However, simulations using published mortality estimates resulted in a rapidly growing population (Supplementary Figure S1). To achieve demographic stability, we increased the mortality rates for individuals aged 0 to 1 years by 6% for both sexes to 42% in males and 41% in females. Actual mortality at <3 months old was likely underestimated in the empirical studies because the initial number of young present in the pouch at birth (one or two) could not readily be established, preventing a reliable estimation of partial litter loss [39]. Accordingly, increasing mortality at age 0–1 years by 6% to achieve demographic stability was deemed a realistic adjustment.

2.8. Sensitivity Analysis

The sensitivity of PVAs to input parameter estimates was investigated using perturbation analysis—using simulations of scenario *1990s Condition* (with no inbreeding depression) while increasing or decreasing a single input parameter by 10% relative to its baseline value (Table 2). The average growth rate (r) was calculated from 200 replicate runs. The growth rate was used to assess model sensitivity because other parameters, such as survival probability and population size, have theoretical limits of zero and/or one, which limits their usefulness for estimating effect size. The relative sensitivity of the model to each parameter was calculated with the following formula: $(r_+ - r_-)/(0.2 \times r_0)$, where r_+ is the mean r with the parameter increased by 10%, r_- is the mean r with the parameter decreased by 10%, r_0 is the mean r using the baseline parameter value, and 0.2 is equal to the perturbation in the parameter value. These estimates indicated whether certain parameters had a disproportionate effect on models, which can make predictions less reliable if the parameter estimates are inaccurate.

2.9. Simulated Scenarios

Five general scenarios were tested, which are outlined in Table 1 and described in detail below. For all scenarios except *1990s Condition*, the starting population was specified using a studbook file (a list of founder individuals and their characteristics in Vortex) that included the age and sex of each individual at Yellingbo in 2019. Relationships among individuals were not provided in the studbook file. Instead, a kinship matrix was also provided for all scenarios except *1990s Condition* to estimate individual inbreeding. For scenarios in which the lowland population was supplemented with highland individuals, the population started with the same set of 20 (10 male and 10 female) Yellingbo founders that were randomly sampled from the studbook containing 35 Yellingbo individuals in 2019. Translocated animals had the same mortality and reproduction rates as the recipient population once added to the population (not including inbreeding depression effects), and all were assumed to survive translocation. For each simulated scenario, the extinction probability was estimated as the proportion of replicate runs that went extinct, with extinction defined as only individuals of one sex remaining; this definition was used in all subsequent simulations. Mean and standard deviation for output parameter estimates were generated using 200 replicate iterations for each scenario. Increasing the number of replicates above 200 did not decrease the standard error of estimates. Simulation outputs included observed heterozygosity, population size (N), growth rate across all years (r), and time to reach $N = 110$.

(i)   1990s Condition: The effect of limited carrying capacity and inbreeding on population size at Yellingbo

The influence of the limited carrying capacity and inbreeding on the lowland population size at Yellingbo since the 1990s was determined. A population of 110 individuals with all pairwise kinships equal to zero was simulated to represent the population in the 1990s, varying the carrying capacity to $K = 120$ and $K = 1000$. These scenarios were run using actual juvenile mortality estimates rather than those increased by 6% to achieve a demographically stable baseline in other scenarios. Simulations at these carrying capacities were run with and without inbreeding depression.

(ii)   2019 Trajectory: The forward projection of the population size of lowland possums at Yellingbo without genetic rescue

The time until extinction of the lowland population at Yellingbo without genetic rescue was estimated with simulations, starting with 35 lowland individuals alive in 2019 in the absence of supplementation and specifying pair-wise kinships, including inbreeding depression and as actual juvenile mortality rates (as for the *1990 Condition* above). These simulations were carried out using two different carrying capacities; $K = 40$, representing the currently degraded habitat conditions, and $K = 120$, representing higher quality habitat present in the 1990s.

(iii)   Demographic Rescue: The effect of numerical population reinforcement on population growth

Scenarios were run to separate the effect of supplementation on population growth (i.e., demographic rescue) from the effect of genetic rescue. The effect of supplementation on population size was estimated by introducing migrants from a donor population genetically similar to the lowland population. The donor population was made genetically similar to that of Yellingbo by setting non-zero kinships between Yellingbo and donor individuals. Kinship values were set at the beginning of each run by randomly sampling from a Poisson distribution with a mean equal to the average pairwise kinship between individuals at Yellingbo (mean = 0.5, the kinship matrix is provided in the Supporting Data). This was also done for pairwise kinships between individuals within the donor population to make it similarly inbred to Yellingbo. Input allele frequencies for this donor population were identical to those of the lowland population.

(iv)   Genetic Rescue: The supplementation of a new lowland population with highland possums

All genetic rescue scenarios were carried out assuming that a new population would be founded using 20 lowland individuals relocated from Yellingbo. To represent the founding of a new population in high-quality habitat, a carrying capacity of 1000 was specified in genetic rescue scenarios. Scenarios were run to determine the minimum supplementation rate required to establish a demographically stable (N changing by <10% over 50 years) population of at least 110 individuals, starting with 20 lowland founders (Table 1). This target size was chosen to recreate the apparent demographic stability observed at Yellingbo in the 1990s and early 2000s [39]. The number of supplemented individuals increased by increments of two (one of each sex) to maintain equal sex ratios during simulations.

For genetic rescue scenarios, a highland population was created using allele frequencies calculated for the Lake Mountain population. These highland individuals were specified as unrelated to each other, and unrelated to lowland individuals, by setting pairwise kinships to zero. This enabled the simulation of genetic rescue by introducing outbred individuals. The size of the highland population was set to 1000 individuals with the mean and standard deviation set to 1 for survival at all life stages to ensure that the population remained sufficiently large for translocation and maintained genetic diversity for the duration of simulations. All other parameters were the same as those used for the lowland population at Yellingbo.

Genetic rescue effects were measured as changes in the mean observed heterozygosity and the mean population size from the first to the last year of simulation, the mean growth rate for all years (r), and the mean time to reach $N=110$. Changes in genetic diversity in the rescued population were modelled by specifying the observed allele frequencies for 1066 loci for the highland population and lowland individuals alive in 2019, and using these to calculate heterozygosity for each year of simulation.

(v)   Genetic Swamping Test: Determining whether a higher supplementation rate risks the replacement of locally unique alleles

Vortex outputs summary statistics for loci only at the end of simulations. Thus, allele frequencies and allele retention probabilities were estimated in a time series; genetic rescue scenarios were repeated with simulations that ran for 10, 20, 30, and 40 years. These values were compared between different supplementation rates to determine whether higher rates of gene flow reduced the retention of locally unique alleles. Whether unique lowland allele retention probabilities and frequencies differed significantly among supplementation rates was determined using aligned rank transformation analysis of variance (ART-ANOVA) in the *R* package ARTool v.0.10.7 [55]. This test is suitable for non-parametric data with paired observations with categorical predictors. Generalised linear models were fit with ARTool with the mean retention probability or frequency as the response, the translocation rate as a predictor, and with data paired by specifying a random intercept for each allele.

Only 23 loci with locally unique lowland alleles were detected (see Section 3.4). Because such a small number of locally unique alleles would make it difficult to predict the effect of the starting allele frequency on retention probability, we used 230 alleles with frequencies replicating those of the 23 observed unique alleles. Frequencies for these unique alleles were set to zero in the highland population. All alleles were assumed to be independently assorting and selectively neutral. The retention probability for each unique lowland allele was given by the proportion of replicate runs in which it did not go extinct.

3. Results

3.1. Sensitivity Analysis

Perturbation analysis indicated that no single input parameter had a greatly disproportionate effect on the population growth rate (Table 3). Annual adult mortality (2+ years) had the greatest relative impact on the population growth rate, closely followed by the litter size (average number of individuals per litter). Point values for r at each parameter value are given in Supplementary Figure S4.

Table 3. Sensitivity analysis output. The relative sensitivity of PVAs to different input parameters based on perturbation analysis.

Input Parameter	Sensitivity Index
0–1 year annual mortality	−0.29
1–2 year annual mortality	−0.11
2+ year annual mortality	−0.44
Proportion of females breeding	0.35
Brood size	0.39

3.2. Trajectory of the Lowland Population without Genetic Management

In the absence of inbreeding depression, simulations using demographic rates estimated in the 1990s showed that population growth was limited by the carrying capacity (Figure 2). A large carrying capacity of 1000 resulted in positive growth ($r = 0.02$, SD = 0.08) for the 50-year simulated timeframe. Limiting the carrying capacity to a more realistic value of 120 resulted in population decline ($r = -0.013$, SD = 0.10) for the duration of the simulations. When inbreeding depression was incorporated into the models that assumed that individuals in the starting population are unrelated (MK = 0), the population declined regardless of whether the carrying capacity was 1000 ($r = -0.028$, SD = 0.12) or 120 ($r = -0.039$, SD = 0.15), although in the former case, it initially increased for half of the simulation period (Figure 2).

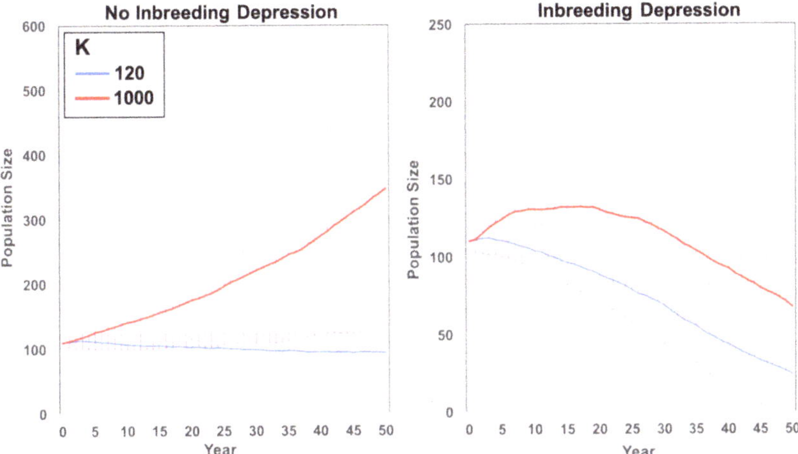

Figure 2. Simulated change in the mean population size ± 1 SD for 200 replicate simulations using mortality estimates for lowland individuals at Yellingbo in the 1990s, with inbreeding depression either excluded or included in models. Simulations were carried out with the carrying capacity (*K*) set to 120 to reflect the limited habitat availability in the reserve, and 1000 to determine whether greater habitat availability could yield a demographically stable population without inbreeding depression. Simulation settings are detailed in Table 1 (*1990s Condition*).

Simulations of the Yellingbo population from 2019 onwards without genetic rescue and with the level of inbreeding informed by genetic estimates (kinship matrix and 2019 allele frequencies) resulted in extinction probabilities of 50% within 11 years, and 100% within 23 years (Figure 3). The mean time to extinction was 11.7 years (SD = 3.4) with a growth rate (*r*) of −0.20 (SD = 0.24) regardless of the carrying capacity (*K* = 40 or *K* = 120). This suggests that the lowland population will likely be extinct within two decades without genetic rescue, even if habitat quality is improved.

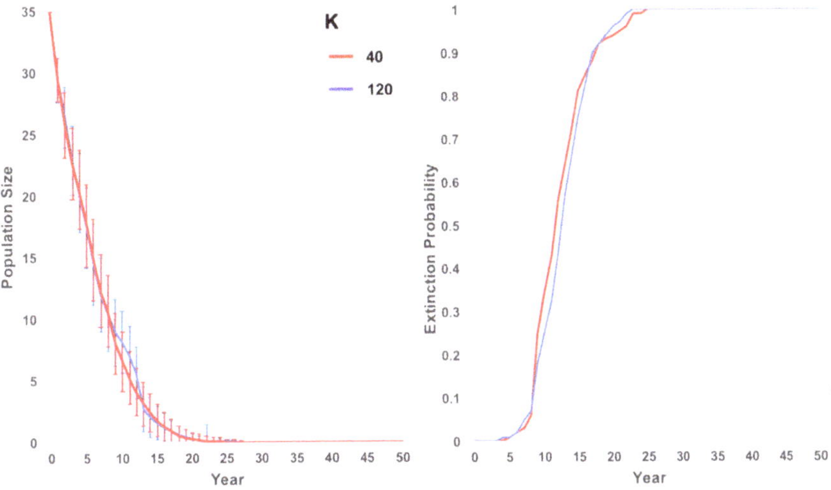

Figure 3. The mean population size (±1 SD) and probability of extinction averaged across 200 replicate simulations, beginning with 35 individuals in Yellingbo with inbreeding depression effects included in the models. The extinction probability is given as the proportion of simulations that had gone extinct by each year, with extinction defined as only individuals of one sex remaining. Simulation settings are given in Table 1 (*2019 Trajectory*).

3.3. Effect of Genetic and Demographic Rescue on Lowland Population Recovery

Mean population size at the end of simulations of genetic rescue scenarios increased with the supplementation rate (Figure 4). Supplementation with 10 highland individuals per year was the only scenario to reach the target size within 10 years, taking an average of 8.27 (SD = 2.0) years with a growth rate of $r = 0.07$ (SD = 0.1). This was closely followed by supplementation with 8 individuals per year, which took 11 years to reach the target size with a growth rate of $r = 0.06$ (SD = 0.09).

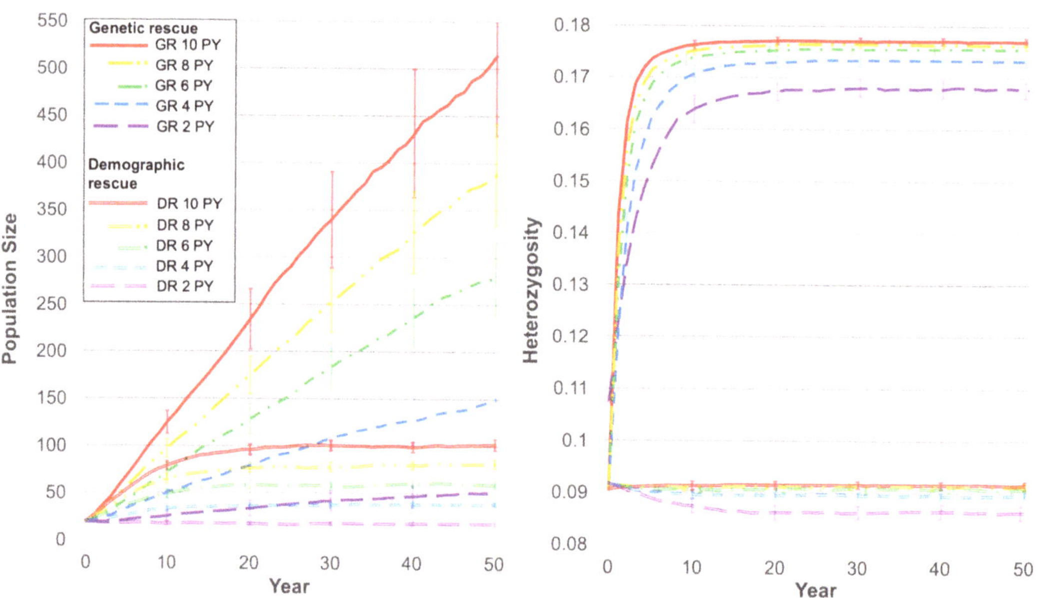

Figure 4. The mean population size ± 1 SD and heterozygosity ± 1 SD for simulations translocating two to ten individuals per year into the inbred lowland population, beginning with 20 lowland founders. Inbreeding depression as included in the models and K=1000. For genetic rescue (GR) scenarios, unrelated individuals were translocated from a simulated highland population. For demographic rescue (DR) scenarios, individuals were introduced with the same allele frequencies and mean pairwise kinships as individuals in the lowland population in 2019. Simulation settings are given in Table 1 (*Genetic Rescue* and *Demographic Rescue*).

Alleviation of inbreeding had an effect distinguishable from merely increasing population size; demographic rescue scenarios (supplementation from the genetically Yellingbo-like population) failed to produce positive population growth in simulations regardless of the supplementation rate (Figure 4). The highest mean population size for the demographic rescue scenarios was 102 individuals and was reached after 49 years at a supplementation rate of 10 individuals per year. Varying the contribution of lethal recessive alleles from 0–75% did not substantially alter the simulation outcomes, although increasing the contribution of lethal alleles unrealistically to 100% resulted in much higher population viability and larger genetic rescue effects (Supplementary Figure S2).

Mean heterozygosity across replicate runs also increased with the supplementation rate for genetic rescue. The highest mean heterozygosity reached in simulations was 0.177 (SD = 0.001), at a supplementation rate of 10 individuals per year. This was the highest value possible, given the genetic diversity of the highland population (heterozygosity = 0.18). Maximum heterozygosity was reached sooner with higher rates of gene flow. For genetic rescue, all translocation rates reached at least 90% of the source population's heterozygosity within ten years. As with the population size, the increase in genetic diversity from genetic rescue was distinguishable from that of demographic rescue. The highest heterozygosity for any demographic rescue scenario was 0.092, which was the highest possible, matching the genetic diversity at Yellingbo.

3.4. Locally Unique Alleles in the Lowland Population at Yellingbo

Out of 4218 loci examined, 23 loci (0.55%) were found to harbour alleles that occurred only in the lowland population. This excluded 70 alleles that occurred in the heterozygous state in a single individual and were removed; including these alleles would increase the proportion of loci with unique lowland alleles to 2.2%. A large proportion of the locally unique alleles were very rare; out of 23 locally unique alleles, 5 (22%), were found at frequencies below 1% in the lowland population (Figure 5a). When all Yellingbo samples from 2011–2019 were considered together, the highest recorded frequency for any unique allele was 57.7%. The highest frequency of any unique allele in 2019 was 19.9%. Of the 23 alleles detected from 2011–2019, 13 (57%) were not detected in 2019 (noting that 14% of individuals were not genotyped in this year).

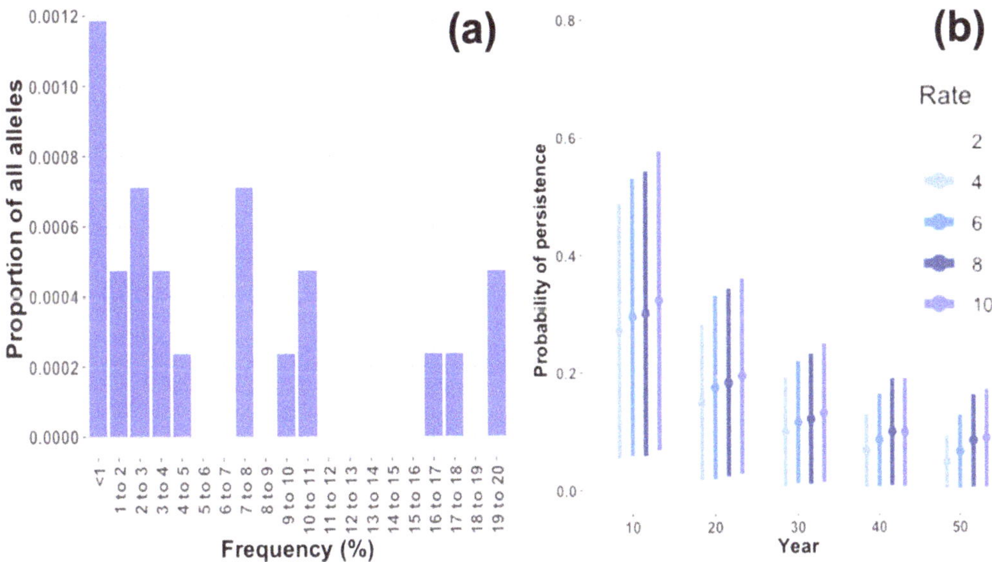

Figure 5. (a) Frequency distribution of locally unique alleles in the lowland population in 2019. Counts of alleles are expressed as a proportion of all alleles in the dataset (*n* = 4218). Frequencies of locally unique alleles from 2011 to 2019 are shown in Supplementary Figure S3. (b) Mean ± 1 SD retention probability for locally unique alleles of any initial frequency during 50 years of simulated genetic rescue. Simulations began with 20 Yellingbo individuals, supplemented with different numbers of highland individuals per year (Rate). Probability and frequency values were output every ten years, with values calculated from 200 replicate runs of each scenario. A carrying capacity of 1000 was specified.

3.5. Risk of Decreased Retention of Locally Unique Alleles from Genetic Rescue

The overall mean retention probability for locally unique alleles was 0.29 (SD = 0.01) at 10 years of genetic rescue, declining to 0.06 (SD = 0.07) at 50 years (Figure 5b). When considered separately, the retention probability for alleles with starting frequencies >10% (Figure 6a) during genetic rescue were somewhat higher, with the highest mean retention probability after 10 years of genetic rescue being 0.68 (SD = 0.11) and the lowest 0.54 (SD = 0.10). After 50 years of genetic rescue, the highest retention probability at any translocation rate was 0.21 (SD = 0.05). For locally unique alleles with starting frequencies <10% (Figure 6b), the highest mean retention probability at 10 years of genetic rescue was much lower at

0.20 (SD = 0.16), with the lowest at 0.15 (SD = 0.12). The highest retention probability at 50 years was 0.04 (SD = 0.04).

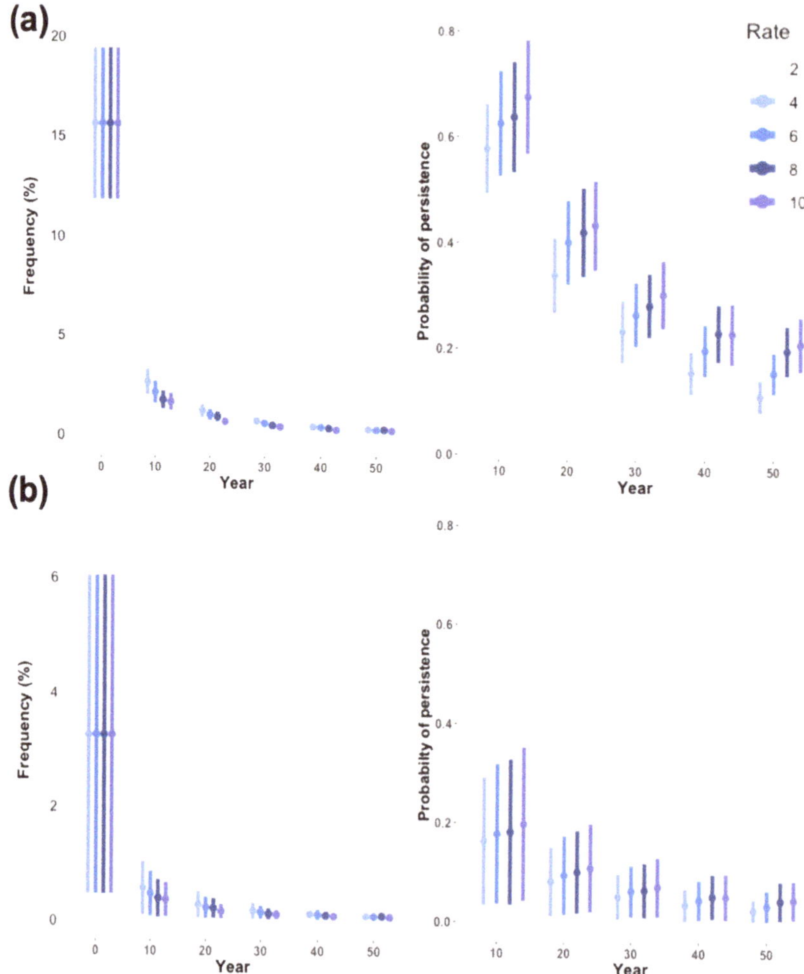

Figure 6. Mean frequencies ± 1 SD and the probability of persistence ± 1 SD of locally unique alleles with starting frequencies of >10% (**a**) and <10% (**b**) in the lowland population. Simulations began with 20 Yellingbo individuals, supplemented with different numbers of highland individuals per year (Rate). Frequency and probability of retention values were output every ten years, with values calculated from 200 replicate runs of each scenario. A carrying capacity of 1000 was specified.

Higher supplementation rates did not produce lower retention probabilities of locally unique alleles during genetic rescue but instead yielded overall higher probabilities (Figure 5). After 20 years of genetic rescue, retention probabilities were substantially higher when the translocation rate was increased from 8 to 10 individuals per year. The difference in retention probability at different translocation rates was statistically significant for all years: 10 years (F = 169.02, df = 4, $p < 0.0001$), 20 years (F = 238.2, df = 4, $p < 0.0001$), 30 years (F = 221.8, df = 4, $p < 0.0001$), 40 years (F = 312.75, df = 4, $p < 0.0001$), and 50 years (F = 393.9, df = 4, $p < 0.0001$). Differences between the translocation rates were also statistically significant for alleles with starting frequencies of <10%: 10 years (F = 64.52, df = 4, $p < 0.0001$),

20 years (F = 212.76, df = 4, $p < 0.0001$), 30 years (F = 115.78, df = 4, $p < 0.0001$), 40 years (F = 157.18, df = 4, $p < 0.0001$), and 50 years (F = 179.6, df = 4, $p < 0.0001$).

Despite higher retention probabilities, locally unique alleles were lower in frequency at higher rates of gene flow (Figure 6). After 10 years of genetic rescue, the highest mean frequency for alleles with a >10% starting frequency was 4.0% (SD = 1.0), which was observed at the lowest supplementation rate of two individuals per year. Using a rate of 10 individuals per year yielded a substantially lower mean frequency of 1.6% (SD = 1.6). For alleles starting at <10% frequency, the highest mean frequency was 0.8% (SD = 0.7), which was also observed at the lowest supplementation rate of the individuals per year. Using a rate of 10 individuals per year yielded a somewhat lower mean frequency of 0.4% (SD = 0.3). The difference in allele frequency at different translocation rates was statistically significant for all years: 10 years (F = 551.21, df = 4, $p < 0.001$), 20 years (F = 268.1, df = 4, $p < 0.0001$), 30 years (F = 143.72, df = 4, $p < 0.0001$), 40 years, (F = 30.5, df = 4, $p < 0.0001$), and 50 years (F = 8.9, df = 4, $p < 0.0001$). Differences between the translocation rates were also statistically significant for the initially rare alleles (<10%): 10 years (F = 64.52, df = 4, $p < 0.001$), 20 years (F = 105.2, df = 4, $p < 0.0001$), 30 years (F = 35.35, df = 4, $p < 0.0001$), 40 years, (F = 8.43, df = 4, $p < 0.0001$), and 50 years (F = 12.31, df = 4, $p < 0.0001$).

4. Discussion

Many threatened wildlife species have highly disjunct distributions with little to no natural gene flow between populations, and translocations are critical for their long-term demographic and genetic stability [21]. Reintroducing populations within the historical range of a species or establishing populations beyond the historical range in low-threat areas (such as predator-free islands or reserves) are increasingly common [30,56,57]. Using predictive models to guide the founding of isolated populations with ongoing translocations is critical to maximising species-wide genetic variation [58]. We demonstrated how PVAs that incorporate inbreeding and inbreeding depression can predict how much locally unique variation can be retained while genetic rescue increases overall genetic variation.

PVAs predicted a very high probability that the lowland population of Leadbeater's possum will go extinct within two decades (100% in 23 years) without any intervention. As environmental impacts on mortality and reproduction were minimised in simulations to make inferences based primarily upon genetics, this is likely an overestimate of time to extinction. Inbreeding depression strongly increased population decline in simulations, although a low carrying capacity was shown to curtail population size, even in the absence of inbreeding depression. This highlights the importance of providing lowland individuals with high-quality habitat to achieve population growth under genetic rescue. Reinforcing the population with only Yellingbo individuals (demographic rescue) did not sufficiently support population growth to reach our target size of 110 individuals, even using the largest translocation rate of 10 individuals per year. We demonstrated that a modest portion of neutral locally unique variation detected in the lowland population can likely be preserved through a combination of population expansion and outcrossing. Approximately 26% of unique lowland alleles occurred at frequencies above 10%, which had a >50% probability of persistence after 10 years of rescue.

There were several key assumptions in our PVAs that could bias estimates of genetic rescue and allele retention. Random mating among highland and lowland individuals was assumed in all scenarios. Preferential mating among individuals originating from similar habitat or populations can impede outcrossing attempts [59]. For Leadbeater's possum, this could cause translocated highland individuals to increase in abundance and supplant lowland individuals, increasing the risk of genetic swamping. This may be prevented by the careful choice of translocation sites, maximising the probability of highland-lowland pairings. Assortative mating among lowland individuals could also subvert genetic rescue efforts. However, females of some species become more selective in their mate choice if they are inbred, developing a preference for outbred males [60]. For lowland possums, this would result in a preference for highland mates, potentially increasing admixture rates and improving locally unique allele retention rates by increasing the fitness of individuals carrying such alleles, as well as increasing population growth and thus, reducing genetic drift. Our PVA models also assumed that the currently observed unique genetic variation is selectively neutral. If locally unique alleles are favourable, selection will increase their likelihood of persistence and frequency more so than if they are neutral. The effect of selection is expected to increase over time with increasing population size [22]. Alternatively, lowland-specific genetic variation may be maladapted if lowland individuals are relocated to a habitat that differs from Yellingbo, in which case, this variation would be selected against. All alleles were also assumed to be unlinked, whereas linkage can promote allele retention by genetic hitch-hiking of neutral alleles linked to loci under selection [31].

PVAs predicted that the supplementation rates of up to 10 individuals per year for 8 years will likely be necessary to meet population recovery targets. Although this supplementation rate is substantial, lower rates and less frequent supplements will likely be sufficient in reality to achieve the target population size. This is because the demographic rates modelled were estimated from data collected from the Yellingbo population during the 1990s, at which time the population was already inbred, having roughly half the genetic diversity of highland populations [36]. Therefore, the positive effects of genetic rescue and immigration will likely be greater than those estimated in this study, and fewer translocations might be necessary to provide the desired population benefits. Furthermore, Vortex does not account for a fixed genetic load in recipient populations [41], which means that fitness could be further improved by increasing heterozygosity at loci with fixed (100% homozygous) deleterious alleles. Using assumptions similarly optimistic as ours, such as completely unrelated migrants, population growth projections were still underestimated by Vortex relative to the real data for translocations in another marsupial species (*Bettongia penicillata ogilbyi*) [61]. Thus, it is conceivable that the required rate of translocation estimated in our study is an overestimate. Conversely, a higher rate of supplementation may be necessary given that all individuals in our simulations were assumed to have the same mortality rate as local individuals. This is unrealistic, given that elevated post-translocation mortality is common in wildlife [62]. Genetic rescue should be implemented within an adaptive management framework, with post-translocation monitoring data incorporated into the models to make updated predictions [63,64]. Ongoing monitoring of fitness will also enable incorporation into future models of unlikely but possible effects of outbreeding depression and maladaptation. The genetic diversity and viability of source populations must also be considered, and further PVAs can help predict and mitigate the negative impacts of harvest [58,65].

The sensitivity of our PVAs to different parameter estimates suggests that management interventions that improve survival and reproductive output could facilitate more rapid population growth during genetic rescue. In particular, the proportion of adult females that were breeding and adult mortality had somewhat larger effects on growth rates than other parameters. Establishing colonies with translocated individuals and unpaired local possums could increase the proportion of females breeding in any given year. Adult mortality could be decreased by feral predator elimination measures, given that predation by feral cats and foxes is a known cause of mortality in Leadbeater's possum [66], and a common cause of translocation failure in Australian mammals [67]. Such measures could reduce the amount of supplementation required to reach a target population size.

Lower retention of locally unique lowland alleles at higher rates of supplementation with highland individuals was not observed in any of the simulated scenarios. However, locally unique alleles did decrease in frequency at higher supplementation rates, which is likely an effect of higher proportions of immigrants diluting neutral allele frequencies. A higher retention probability under more intensive gene flow was presumably caused by weaker genetic drift for scenarios where the population size increased more quickly. Thus, under the conditions used in simulations for the lowland population of Leadbeater's possum, genetic drift poses a greater threat to the preservation of unique genetic variation than dilution at higher rates of translocation. The number of lowland and highland founders will be a strong determinant of the number of alleles retained. Based on allele retention models in another translocated marsupial, using 20 founders is expected to retain ~60% of rare (frequency = 0.05) alleles and using at least 60 founders is expected to retain ~90% of rare alleles [30]. Minimising genetic and demographic stochasticity in small populations by rapid supplementation has been recommended in other species to maximise the effectiveness of population reinforcement [68]. Our study suggests that this principle could also apply to genetic rescue and the preservation of locally unique variation. Although the frequency of these locally unique alleles may be reduced at higher supplementation rates, this would be offset by the benefits of a larger population size (such as reduced stochasticity), and selection will increase their frequency if they present a fitness advantage.

Previous experimental investigations into genetic swamping effects following genetic rescue have focused on locally adaptive alleles [22,69]. However, genetic variation that is not currently subject to selection (such as cryptic variation that only alters phenotype under atypical circumstances) can become useful if environments change; such enhancement of evolutionary potential would preserve apparently neutral variation in the face of uncertain futures [14,70]. Although only a small proportion of alleles (0.55–2.2% of loci) were unique to the lowland population at Yellingbo and none were fixed, it is possible that they could contribute to local adaptation or evolutionary potential. Much unique variation will have been lost from the lowland population owing to the small population size over decades, with an observable reduction in the number of locally unique alleles from 2011 to 2019 (Supplementary Figure S4). Nonetheless, locally adaptive genetic variation can persist in the presence of strong genetic drift [71]. The detection of locally unique variation is also dependent on marker type and density, with SNPs representing only one form of genetic variation. Other types of variation that contribute to local adaptation include structural and copy number variants [72]. Such variation is unlikely represented in SNP datasets, making the amount of unique variation among populations an underestimate when using SNPs alone. Deleterious alleles that are recent mutations tend to be prevalent at low frequencies [73], and many locally unique alleles detected were very rare (17% with <1% frequency). However, it is unclear whether the locally unique alleles detected could be deleterious, because intense genetic drift in the lowland population will have strongly influenced allele frequencies.

The contribution of ongoing post-rescue gene flow to demographic and genetic stability is another important management consideration. Although there is some potential that supplementing the lowland population with individuals from large populations will increase the genetic load, maintaining heterozygosity and minimising inbreeding via occasional gene flow will mask recessive genetic load. Some studies have cited pronounced population declines following the immigration of as few as a single individual as evidence that genetic rescue can further imperil small populations by introducing deleterious alleles that are more numerous in larger populations [74]. Fitness decline with insufficient genetic rescue is unsurprising because inbreeding (identity-by-descent) inevitably increases to pre-translocation levels in very small populations without ongoing gene flow [75]. Founding a new lowland population near existing Central Highlands populations could enable dispersal between populations, which would increase long-term population viability. Natural dispersal could also facilitate the spread of lowland genetic variation that is potentially adaptive to warmer conditions, improving the fitness of Central Highlands populations under climate change. The establishment of self-sustaining populations is a highly sought-after goal in conservation management [76]. However, populations that are genetically isolated with $N_e < 1000$ will inevitably require at least occasional assisted gene flow if natural gene flow is not possible [1,58]. Early interventions that prevent extreme reductions in N_e and mitigating the underlying causes of a small population size are the best means of reducing dependency on conservation management.

5. Conclusions

We estimate that without intervention, all unique lowland Leadbeater's possum genetic variation will be lost due to extinction of the last population of the lineage within approximately two decades, whereas modest amounts of this unique variation can be preserved with genetic rescue. A higher proportion of locally unique variation is predicted to be retained by adding a greater number of highland animals each year with genetic rescue of the lowland population. Ongoing gene flow from highland populations will also be critical to maintaining long-term demographic and genetic stability within the lowland population. Although PVAs are based on many assumptions, they can form an integral component of adaptive genetic management and offer flexibility to incorporate post-translocation data to resolve uncertainties such as relative fitness among outbred and inbred individuals.

Supplementary Materials: The following are available online at https://www.mdpi.com/article/10.3390/d13080382/s1, Figure S1. Population growth using values of mortality at 0–1 years estimated in the 1990's and using estimated values +6% to produce a demographically stable population in the absence of inbreeding depression. Figure S2. Effect of percent inbreeding depression due to lethal recessive alleles (as opposed to heterozygote advantage) on genetic rescue. Effects are shown as change in population size (N) with genetic rescue by supplementing Yellingbo founders with two highland individuals per year. Percentages of 0–75 inbreeding due to lethal recessives produced comparable changes in population size. Inbreeding depression 100% due to lethals produced much more positive population growth. Figure S3. Frequencies of locally-unique alleles at Yellingbo from 2011–2019, and 2019 alone. Figure S4. PVA input parameter sensitivity tests. Mean growth rate (r) from 200 replicate runs of 50-year simulations. All scenarios began with 110 individuals, excluded inbreeding depression, and used a carrying capacity of 1,0000. A baseline scenario was run using actual input parameter values (except with 0–1 yr mortality increased by 6%), which gave a stable population of approximately 110 individuals. Output values for this baseline scenario are shown by the dashed line. For each scenario shown on the x-axis, a single input parameter was either increased (Low) or increased (High) by 10% of its baseline value to determine its relative influence on simulation outcomes.

Author Contributions: Conceptualization, J.P.Z. and P.S.; methodology, J.P.Z., D.H., A.P. and P.S.; software, J.P.Z.; validation, J.P.Z., D.H., A.P. and P.S.; formal analysis, J.P.Z.; investigation, J.P.Z.; resources, P.S. and J.P.Z.; data curation, J.P.Z.; writing—original draft preparation, J.P.Z.; writing— review and editing, J.P.Z., D.H., A.P. and P.S.; visualization, J.P.Z.; supervision, P.S and A.P.; project administration, J.P.Z. and P.S.; funding acquisition, P.S. and J.P.Z. All authors have read and agreed to the published version of the manuscript.

Funding: This research was funded by an ARC Linkage Grant LP160100482 with Partner Organizations Victorian Department of Environment, Land, Water, and Planning (DELWP), Diversity Arrays Technology, Zoos Victoria, Australian Capital Territory Environment, Planning & Sustainable Development Directorate, University of Canberra, and Western Australian Department of Biodiversity, Conservation and Attractions. Joanne Antrobus from Parks Victoria has provided substantial support to the long-term monitoring program for Leadbeater's possum at Yellingbo. J.P.Z. and D.H. were supported by Monash University internal funds and Research and Training Program Scholarships (formerly Australian Postgraduate Award), and awards from the Holsworth Wildlife Research Endowment. This research drew on materials from an Australian Academy of Sciences Research Award for the Conservation of Endangered Australian Vertebrate Species to Andrea Taylor.

Institutional Review Board Statement: All population monitoring and tissue sample collection procedures were approved by the Zoos Victoria Animal Ethics Committee (project ZV14004 approved 16/04/2014 and project ZV18001 approved 15/03/2018).

Informed Consent Statement: Not applicable.

Data Availability Statement: Supporting data for this publication are publicly available at the Monash University online data repository (https://doi.org/10.26180/15019797, accessed on 12 August 2021).

Acknowledgments: We wish to thank Bob Lacy for extremely useful advice regarding PVAs. We thank the editors and *Diversity* for the invitation to submit a manuscript for consideration in this special issue, three anonymous reviewers for their comments, and Carolyn Hogg and Steve Cooper who assessed an earlier version.

Conflicts of Interest: The authors declare no conflict of interest.

References

1. Frankham, R.; Ballou, J.D.; Ralls, K.; Eldridge, M.D.B.; Dudash, M.R.; Fenster, C.B.; Lacy, R.C.; Sunnucks, P. *Genetic Management of Fragmented Animal and Plant Populations*; Oxford University Press: Oxford, UK, 2017.

2. Whiteley, A.R.; Fitzpatrick, S.W.; Funk, W.C.; Tallmon, D.A. Genetic Rescue to the Rescue. *Trends Ecol. Evol.* **2015**, *30*, 42–49. [CrossRef] [PubMed]

3. Hedrick, P.W.; Garcia-Dorado, A. Understanding Inbreeding Depression, Purging, and Genetic Rescue. *Trends Ecol. Evol.* **2016**, *31*, 940–952. [CrossRef]

4. Hedrick, P.W.; Fredrickson, R. Genetic Rescue Guidelines with Examples from Mexican Wolves and Florida Panthers. *Conserv. Genet.* **2010**, *11*, 615–626. [CrossRef]

5. Eizaguirre, C.; Baltazar-Soares, M. Evolutionary Conservation–Evaluating the Adaptive Potential of Species. *Evol. Appl.* **2014**, *7*, 963–967. [CrossRef]

6. Fagan, W.F.; Holmes, E.E. Quantifying the Extinction Vortex. *Ecol. Lett.* **2006**, *9*, 51–60. [CrossRef] [PubMed]

7. Blomqvist, D.; Pauliny, A.; Larsson, M.; Flodin, L.-A. Trapped in the Extinction Vortex? Strong Genetic Effects in a Declining Vertebrate Population. *BMC Evol. Biol.* **2010**, *10*, 33. [CrossRef]

8. vander Wal, E.; Garant, D.; Calmé, S.; Chapman, C.A.; Festa-Bianchet, M.; Millien, V.; Rioux-Paquette, S.; Pelletier, F. Applying Evolutionary Concepts to Wildlife Disease Ecology and Management. *Evol. Appl.* **2014**, *7*, 856–868. [CrossRef] [PubMed]

9. Hoffmann, A.A.; Miller, A.D.; Weeks, A.R. Genetic Mixing for Population Management: From Genetic Rescue to Provenancing. *Evol. Appl.* **2021**, *14*, 634–652. [CrossRef] [PubMed]

10. Frankham, R. Genetic Rescue of Small Inbred Populations: Meta-Analysis Reveals Large and Consistent Benefits of Gene Flow. *Mol. Ecol.* **2015**, *24*, 2610–2618. [CrossRef] [PubMed]

11. Love Stowell, S.M.; Pinzone, C.A.; Martin, A.P. Overcoming Barriers to Active Interventions for Genetic Diversity. *Biodivers. Conserv.* **2017**, *26*, 1753–1765. [CrossRef]

12. Odell, E.A.; Heffelfinger, J.R.; Rosenstock, S.S.; Bishop, C.J.; Liley, S.; González-Bernal, A.; Velasco, J.A.; Martínez-Meyer, E. Perils of Recovering the Mexican Wolf Outside of Its Historical Range. *Biol. Conserv.* **2018**, *220*, 290–298. [CrossRef]

13. Bell, D.A.; Robinson, Z.L.; Funk, W.C.; Fitzpatrick, S.W.; Allendorf, F.W.; Tallmon, D.A.; Whiteley, A.R. The Exciting Potential and Remaining Uncertainties of Genetic Rescue. *Trends Ecol. Evol.* **2019**, *34*, 1070–1079. [CrossRef]

14. Harrisson, K.A.; Pavlova, A.; Telonis-Scott, M.; Sunnucks, P. Using Genomics to Characterize Evolutionary Potential for Conservation of Wild Populations. *Evol. Appl.* **2014**, *7*, 1008–1025. [CrossRef]

15. Roberts, D.G.; Gray, C.A.; West, R.J.; Ayre, D.J. Marine Genetic Swamping: Hybrids Replace an Obligately Estuarine Fish. *Mol. Ecol.* **2010**, *19*, 508–520. [CrossRef]

16. Hedrick, P.W. Gene Flow and Genetic Restoration: The Florida Panther as a Case Study. *Conserv. Biol.* **1995**, *9*, 996–1007. [CrossRef] [PubMed]

17. Liddell, E.; Sunnucks, P.; Cook, C.N. To Mix or Not to Mix Gene Pools for Threatened Species Management? Few Studies Use Genetic Data to Examine the Risks of Both Actions, but Failing to Do so Leads Disproportionately to Recommendations for Separate Management. *Biol. Conserv.* **2021**, *256*, 109072. [CrossRef]

18. Aitken, S.N.; Whitlock, M.C. Assisted Gene Flow to Facilitate Local Adaptation to Climate Change. *Annu. Rev. Ecol. Evol. Syst.* **2013**, *44*, 367–388. [CrossRef]

19. Sjöstrand, A.E.; Sjödin, P.; Jakobsson, M. Private Haplotypes Can Reveal Local Adaptation. *BMC Genet.* **2014**, *15*, 61. [CrossRef] [PubMed]

20. Thrimawithana, A.H.; Ortiz-Catedral, L.; Rodrigo, A.; Hauber, M.E. Reduced Total Genetic Diversity Following Translocations? A Metapopulation Approach. *Conserv. Genet.* **2013**, *14*, 1043–1055. [CrossRef]

21. Weeks, A.R.; Stoklosa, J.; Hoffmann, A.A. Conservation of Genetic Uniqueness of Populations May Increase Extinction Likelihood of Endangered Species: The Case of Australian Mammals. *Front. Zool.* **2016**, *13*, 31. [CrossRef]

22. Fitzpatrick, S.W.; Bradburd, G.S.; Kremer, C.T.; Salerno, P.E.; Angeloni, L.M.; Funk, W.C. Genomic and Fitness Consequences of Genetic Rescue in Wild Populations. *Curr. Biol.* **2020**, *30*, 517–522.e5. [CrossRef]

23. Willi, Y.; Griffin, P.; van Buskirk, J. Drift Load in Populations of Small Size and Low Density. *Heredity* **2013**, *110*, 296–302. [CrossRef]

24. Fernandez-Fournier, P.; Lewthwaite, J.M.M.; Mooers, A.Ø. Do We Need to Identify Adaptive Genetic Variation When Prioritizing Populations for Conservation? *Conserv. Genet.* **2021**, *22*, 205–216. [CrossRef]

25. Ørsted, M.; Hoffmann, A.A.; Sverrisdóttir, E.; Nielsen, K.L.; Kristensen, T.N. Genomic Variation Predicts Adaptive Evolutionary Responses Better than Population Bottleneck History. *PLoS Genet.* **2019**, *15*, e1008205. [CrossRef] [PubMed]

26. Hoban, S.; Bertorelle, G.; Gaggiotti, O.E. Computer Simulations: Tools for Population and Evolutionary Genetics. *Nat. Rev. Genet.* **2012**, *13*, 110–122. [CrossRef]

27. Pavlova, A.; Beheregaray, L.B.; Coleman, R.; Gilligan, D.; Harrisson, K.A.; Ingram, B.A.; Kearns, J.; Lamb, A.M.; Lintermans, M.; Lyon, J.; et al. Severe Consequences of Habitat Fragmentation on Genetic Diversity of an Endangered Australian Freshwater Fish: A Call for Assisted Gene Flow. *Evol. Appl.* **2017**, *10*, 531–550. [CrossRef]
28. Reynolds, M.H.; Weiser, E.; Jamieson, I.; Hatfield, J.S. Demographic Variation, Reintroduction, and Persistence of an Island Duck (Anas Laysanensis). *J. Wildl. Manag.* **2013**, *77*, 1094–1103. [CrossRef]
29. Schueller, A.M.; Hayes, D.B. Minimum Viable Population Size for Lake Sturgeon (Acipenser Fulvescens) Using an Individual-Based Model of Demographics and Genetics. *Can. J. Fish. Aquat. Sci.* **2010**, *68*, 62–73. [CrossRef]
30. White, D.J.; Ottewell, K.; Spencer, P.B.S.; Smith, M.; Short, J.; Sims, C.; Mitchell, N.J. Genetic Consequences of Multiple Translocations of the Banded Hare-Wallaby in Western Australia. *Diversity* **2020**, *12*, 448. [CrossRef]
31. Harris, K.; Zhang, Y.; Nielsen, R. Genetic Rescue and the Maintenance of Native Ancestry. *Conserv. Genet.* **2019**, *20*, 59–64. [CrossRef]
32. Greet, J.; Harley, D.; Ashman, K.; Watchorn, D.; Duncan, D. The Vegetation Structure and Condition of Contracting Lowland Habitat for Leadbeater's Possum (Gymnobelideus Leadbeateri). *Aust. Mammal.* **2020**. First published online 16 December 2020. [CrossRef]
33. Harley, D. An Overview of Actions to Conserve Leadbeater's Possum "Gymnobelideus Leadbeateri". *Vic. Nat.* **2016**, *133*, 85–97. [CrossRef]
34. Harley, D.; Worley, M.; Harley, T. The Distribution and Abundance of Leadbeater's Possum Gymnobelideus Leadbeateri in Lowland Swamp Forest at Yellingbo Nature Conservation Reserve. *Aust. Mammal.* **2005**, *27*, 7–15. [CrossRef]
35. Hansen, B.D.; Taylor, A.C. Isolated Remnant or Recent Introduction? Estimating the Provenance of Yellingbo Leadbeater's Possums by Genetic Analysis and Bottleneck Simulation. *Mol. Ecol.* **2008**, *17*, 4039–4052. [CrossRef]
36. Hansen, B.D.; Harley, D.K.P.; Lindenmayer, D.B.; Taylor, A.C. Population Genetic Analysis Reveals a Long-Term Decline of a Threatened Endemic Australian Marsupial. *Mol. Ecol.* **2009**, *18*, 3346–3362. [CrossRef]
37. Zilko, J.P.; Harley, D.; Hansen, B.; Pavlova, A.; Sunnucks, P. Accounting for Cryptic Population Substructure Enhances Detection of Inbreeding Depression with Genomic Inbreeding Coefficients: An Example from a Critically Endangered Marsupial. *Mol. Ecol.* **2020**, *29*, 2978–2993. [CrossRef]
38. Harley, D. The Application of Zoos Victoria's "Fighting Extinction" Commitment to the Conservation of Leadbeater's Possum "Gymnobelideus Leadbeateri". *Vic. Nat.* **2012**, *129*, 175–180. [CrossRef]
39. Harley, D.K.P.; Lill, A. Reproduction in a Population of the Endangered Leadbeater's Possum Inhabiting Lowland Swamp Forest. *J. Zool.* **2007**, *272*, 451–457. [CrossRef]
40. Lacy, R.C. Lessons from 30 Years of Population Viability Analysis of Wildlife Populations. *Zoo Biol.* **2019**, *38*, 67–77. [CrossRef]
41. Lacy, R.C. VORTEX: A Computer Simulation Model for Population Viability Analysis. *Wildl. Res.* **1993**, *20*, 45–65. [CrossRef]
42. O'Grady, J.J.; Brook, B.W.; Reed, D.H.; Ballou, J.D.; Tonkyn, D.W.; Frankham, R. Realistic Levels of Inbreeding Depression Strongly Affect Extinction Risk in Wild Populations. *Biol. Conserv.* **2006**, *133*, 42–51. [CrossRef]
43. Simmons, M.J.; Crow, J.F. Mutations Affecting Fitness in Drosophila Populations. *Annu. Rev. Genet.* **1977**, *11*, 49–78. [CrossRef] [PubMed]
44. Harley, D.K.P. The Life History and Conservation of Leadbeater's Possum (Gymnobelideus leadbeateri) in Lowland Swamp Forest. PhD Thesis, Monash University, Melbourne, Australia, 2005.
45. Charlesworth, D.; Willis, J.H. The Genetics of Inbreeding Depression. *Nat. Reviews. Genet.* **2009**, *10*, 783–796. [CrossRef] [PubMed]
46. Sunnucks, P.; Hales, D.F. Numerous Transposed Sequences of Mitochondrial Cytochrome Oxidase I-II in Aphids of the Genus Sitobion (Hemiptera: Aphididae). *Mol. Biol. Evol.* **1996**, *13*, 510–524. [CrossRef]
47. Kilian, A.; Wenzl, P.; Huttner, E.; Carling, J.; Xia, L.; Blois, H.; Caig, V.; Heller-Uszynska, K.; Jaccoud, D.; Hopper, C.; et al. Diversity Arrays Technology: A Generic Genome Profiling Technology on Open Platforms. In *Methods in Molecular Biology*; Humana Press: Clifton, NJ, USA, 2012; Volume 888, pp. 67–89.
48. Gruber, B.; Unmack, P.J.; Berry, O.F.; Georges, A. Dartr: An r Package to Facilitate Analysis of SNP Data Generated from Reduced Representation Genome Sequencing. *Mol. Ecol. Resour.* **2018**, *18*, 691–699. [CrossRef]
49. Kamvar, Z.N.; Tabima, J.F.; Grünwald, N.J. Poppr: An R Package for Genetic Analysis of Populations with Clonal, Partially Clonal, and/or Sexual Reproduction. *PeerJ* **2014**, *2*, e281. [CrossRef]
50. Thompson, E.A. Population Correlation and Population Kinship. *Theor. Popul. Biol.* **1976**, *10*, 205–226. [CrossRef]
51. Städele, V.; Vigilant, L. Strategies for Determining Kinship in Wild Populations Using Genetic Data. *Ecol. Evol.* **2016**, *6*, 6107–6120. [CrossRef]
52. Goudet, J. Hierfstat, a Package for r to Compute and Test Hierarchical F-Statistics. *Mol. Ecol. Notes* **2005**, *5*, 184–186. [CrossRef]
53. Goudet, J.; Kay, T.; Weir, B.S. How to Estimate Kinship. *Mol. Ecol.* **2018**, *27*, 4121–4135. [CrossRef] [PubMed]
54. Weir, B.S.; Goudet, J. A Unified Characterization of Population Structure and Relatedness. *Genetics* **2017**, *206*, 2085. [CrossRef] [PubMed]
55. Wobbrock, J.O.; Findlater, L.; Gergle, D.; Higgins, J.J. The Aligned Rank Transform for Nonparametric Factorial Analyses Using Only Anova Procedures. In Proceedings of the SIGCHI Conference on Human Factors in Computing Systems; Association for Computing Machinery, New York, NY, USA, 12 May 2011; pp. 143–146.
56. Ryan, C.M.; Hobbs, R.J.; Valentine, L.E. Bioturbation by a Reintroduced Digging Mammal Reduces Fuel Loads in an Urban Reserve. *Ecol. Appl.* **2020**, *30*, e02018. [CrossRef] [PubMed]

57. Burbidge, A.A.; Abbott, I. Mammals on Western Australian Islands: Occurrence and Preliminary Analysis. *Aust. J. Zool.* **2017**, *65*, 183–195. [CrossRef]

58. Pacioni, C.; Wayne, A.F.; Page, M. Guidelines for Genetic Management in Mammal Translocation Programs. *Biol. Conserv.* **2019**, *237*, 105–113. [CrossRef]

59. Slade, B.; Parrott, M.L.; Paproth, A.; Magrath, M.J.L.; Gillespie, G.R.; Jessop, T.S. Assortative Mating among Animals of Captive and Wild Origin Following Experimental Conservation Releases. *Biol. Lett.* **2014**, *10*, 20140656. [CrossRef] [PubMed]

60. Pilakouta, N.; Smiseth, P.T. Female Mating Preferences for Outbred versus Inbred Males Are Conditional upon the Female's Own Inbreeding Status. *Anim. Behav.* **2017**, *123*, 369–374. [CrossRef]

61. Pacioni, C.; Atkinson, A.; Trocini, S.; Rafferty, C.; Morley, K.; Spencer, P.B.S. Is Supplementation an Efficient Management Action to Increase Genetic Diversity in Translocated Populations? *Ecol. Manag. Restor.* **2020**, *21*, 123–130. [CrossRef]

62. Armstrong, D.P.; Seddon, P.J. Directions in Reintroduction Biology. *Trends Ecol. Evol.* **2008**, *23*, 20–25. [CrossRef]

63. Westgate, M.J.; Likens, G.E.; Lindenmayer, D.B. Adaptive Management of Biological Systems: A Review. *Biol. Conserv.* **2013**, *158*, 128–139. [CrossRef]

64. Pacioni, C.; Trocini, S.; Wayne, A.F.; Rafferty, C.; Page, M. Integrating Population Genetics in an Adaptive Management Framework to Inform Management Strategies. *Biodivers. Conserv.* **2020**, *29*, 947–966. [CrossRef]

65. Verdon, S.J.; Mitchell, W.F.; Clarke, M.F. Can Flexible Timing of Harvest for Translocation Reduce the Impact on Fluctuating Source Populations? *Wildl. Res.* **2021**, *28*, 458–469. [CrossRef]

66. McComb, L.B.; Lentini, P.E.; Harley, D.K.P.; Lumsden, L.F.; Antrobus, J.S.; Eyre, A.C.; Briscoe, N.J. Feral Cat Predation on Leadbeater's Possum (Gymnobelideus Leadbeateri) and Observations of Arboreal Hunting at Nest Boxes. *Aust. Mammal.* **2019**, *41*, 262–265. [CrossRef]

67. Pacioni, C.; Williams, M.R.; Lacy, R.C.; Spencer, P.B.S.; Wayne, A.F. Predators and Genetic Fitness: Key Threatening Factors for the Conservation of a Bettong Species. *Pac. Conserv. Biol.* **2017**, *23*, 200–212. [CrossRef]

68. Hardy, M.A.; Hull, S.D.; Zuckerberg, B. Swift Action Increases the Success of Population Reinforcement for a Declining Prairie Grouse. *Ecol. Evol.* **2018**, *8*, 1906–1917. [CrossRef] [PubMed]

69. Poirier, M.-A.; Coltman, D.W.; Pelletier, F.; Jorgenson, J.; Festa-Bianchet, M. Genetic Decline, Restoration and Rescue of an Isolated Ungulate Population. *Evol. Appl.* **2019**, *12*, 1318–1328. [CrossRef]

70. Zheng, J.; Payne, J.L.; Wagner, A. Cryptic Genetic Variation Accelerates Evolution by Opening Access to Diverse Adaptive Peaks. *Science* **2019**, *365*, 347. [CrossRef] [PubMed]

71. Funk, W.C.; Lovich, R.E.; Hohenlohe, P.A.; Hofman, C.A.; Morrison, S.A.; Sillett, T.S.; Ghalambor, C.K.; Maldonado, J.E.; Rick, T.C.; Day, M.D.; et al. Adaptive Divergence despite Strong Genetic Drift: Genomic Analysis of the Evolutionary Mechanisms Causing Genetic Differentiation in the Island Fox (Urocyon Littoralis). *Mol. Ecol.* **2016**, *25*, 2176–2194. [CrossRef]

72. Tigano, A.; Friesen, V.L. Genomics of Local Adaptation with Gene Flow. *Mol. Ecol.* **2016**, *25*, 2144–2164. [CrossRef] [PubMed]

73. Bosse, M.; Megens, H.-J.; Derks, M.F.L.; de Cara, Á.M.R.; Groenen, M.A.M. Deleterious Alleles in the Context of Domestication, Inbreeding, and Selection. *Evol. Appl.* **2018**, *12*, 6–17. [CrossRef] [PubMed]

74. Kyriazis, C.C.; Wayne, R.K.; Lohmueller, K.E. Strongly Deleterious Mutations Are a Primary Determinant of Extinction Risk Due to Inbreeding Depression. *Evol. Lett.* **2021**, *5*, 33–47. [CrossRef]

75. Ralls, K.; Sunnucks, P.; Lacy, R.C.; Frankham, R. Genetic Rescue: A Critique of the Evidence Supports Maximizing Genetic Diversity Rather than Minimizing the Introduction of Putatively Harmful Genetic Variation. *Biol. Conserv.* **2020**, *251*, 108784. [CrossRef]

76. Mussmann, S.M.; Douglas, M.R.; Anthonysamy, W.J.B.; Davis, M.A.; Simpson, S.A.; Louis, W.; Douglas, M.E. Genetic Rescue, the Greater Prairie Chicken and the Problem of Conservation Reliance in the Anthropocene. *R. Soc. Open Sci.* **2021**, *4*, 160736. [CrossRef]

diversity

MDPI

Article

Genetic Diversity and Population Structure of Mesoamerican Scarlet Macaws in an Ex Situ Breeding Population in Mexico

Patricia Escalante-Pliego [1,*], Noemí Matías-Ferrer [2], Patricia Rosas-Escobar [1], Gabriela Lara-Martínez [3], Karol Sepúlveda-González [3] and Rodolfo Raigoza-Figueras [3]

[1] Instituto de Biología, Universidad Nacional Autónoma de México, Ciudad Universitaria, Coyoacán, Mexico City 04510, Mexico; patricia_rosas@yahoo.com

[2] Red de Interacciones Multitróficas, Instituto de Ecología, A. C., Carretera Antigua a Coatepec No. 351, Xalapa 91070, Mexico; noemi.matias@inecol.mx

[3] Xcaret, Carretera Chetumal Puerto Juárez, Km 282, Municipio de Solidaridad, Playa del Carmen 77710, Mexico; glara@xcaret.com (G.L.-M.); ksepulveda@xcaret.com (K.S.-G.); rraigoza@xcaret.com (R.R.-F.)

* Correspondence: tilmatura@ib.unam.mx; Tel.: +52-5554002995

Abstract: Given the interest in the conservation of the Mesoamerican scarlet macaw (*Ara macao cyanoptera*), the Xcaret Park formed an initial reproductive population about 30 years ago, which has progressively grown to a considerable population in captivity. In this work, we focus on the evaluation of the genetic diversity of the captive population, taking two groups into account: its founding (49) and the current breeding individuals (166). The genetic analysis consisted of genotyping six nuclear microsatellite loci that are characterized by their high variability. Tests for all loci revealed a Hardy–Weinberg equilibrium in four loci of the founders and in no loci of the breeding groups. The results showed that the genetic variation in the Xcaret population was relatively high (founders He = 0.715 SE = 0.074, breeding pairs He = 0.763 SE = 0.050), with an average polymorphism of 7.5 (4–10) alleles per locus in founders and 8.3 (4–14) in breeding pairs. No significant differences in the evaluated genetic diversity indexes were found between both groups. This indicates that the genetic variability in Xcaret has been maintained, probably due to the high number of pairs and the reproductive management strategy. Bayesian analysis revealed five different genetic lineages present in different proportions in the founders and in the breeding pairs, but no population structure was observed between founders and breeding individuals. The analyzed captive individuals showed levels of genetic diversity comparable to reported values from *Ara macao* wild populations. These data indicate that the captive population has maintained a similar genetic diversity as the metapopulation in the Mayan Forest and is an important resource for reintroduction projects, some of which began more than five years ago and are still underway.

Keywords: ex situ conservation; Psittacidae; *Ara macao*; conservation genetics; Xcaret; captive breeding

Citation: Escalante-Pliego, P.; Matías-Ferrer, N.; Rosas-Escobar, P.; Lara-Martínez, G.; Sepúlveda-González, K.; Raigoza-Figueras, R. Genetic Diversity and Population Structure of Mesoamerican Scarlet Macaws in an Ex Situ Breeding Population in Mexico. *Diversity* 2022, 14, 54. https://doi.org/10.3390/d14010054

Academic Editors: Miguel Ferrer, Kym Ottewell and Margaret Byrne

Received: 27 September 2021
Accepted: 5 January 2022
Published: 14 January 2022

1. Introduction

A recommendation made by the Species Survival Committee of the International Union for the Conservation of Nature [1] concerning reintroduction projects mentions the need to include genetic studies, since it is important to try to introduce sufficient genetic variability in the founding individuals of a new population to avoid bottlenecks, greater inbreeding, and possible problems of local adaptation to diseases or environmental changes [2,3]. Genetics is therefore an important aspect in the conservation or recovery program of any species, though by no means the only one [4].

The Mesoamerican scarlet macaw (*Ara macao cyanoptera*) is classified as endangered in Mexico [5] because it has disappeared from most of its original distribution, which used to extend from Tamaulipas through Veracruz, Oaxaca, Tabasco and Chiapas, and as far south as Costa Rica [6–8]. The IUCN received a proposal to consider this subspecies as endangered [9]. The drastic decline in its populations is caused by the poaching of nestlings for the pet trade and loss of its natural habitat: the high evergreen forest [10]. In Mexico, only small remnants are left of what were once abundant populations. Experts estimate that this subspecies may be lost forever if no preservation action is taken in the next 10 years [11]. At present, we can do more for the conservation of this subspecies by reintroducing it in areas where it can be viable in the mid and long term with the use of captive breeding.

Wiedenfeld [6] found that wild populations of scarlet macaws distributed from Mexico to Central America and to the Amazon River basin may in fact be divided into two subspecies, one distributed from Costa Rica to southern Brazil *(Ara m. macao)* and the other from Mexico to Costa Rica *(Ara m. cyanoptera)*. García-Feria [12] analyzed the genetic variation of four contemporary populations of *A. m. cyanoptera* in Chiapas (Mexico), Guatemala, Belize, and Honduras, using two fragments of mitochondrial DNA and a nuclear gene. His conclusions were that there is no genetic break between the studied populations, and that they comprise a cohesive reproductive unit. He observed that 91% of genetic variation was found within populations and only 9% between populations. A more complete phylogeographic study of *Ara macao* using sequences of mitochondrial genes, based on museum specimens, confirmed the existence of two lineages, which must indeed be recognized as subspecies [13]. Their findings in relation to the Mexican and Central American populations indicated that populations from separate sites from the Isthmus of Tehuantepec to the Caribbean slope in Belize and Guatemala did not present any significant substructure at separate sites, forming a single demographic unit or panmictic population, the Mayan Forest metapopulation (Isthmus Tehuantepec-Lancadona-Guatemala-Belize), but found a second demographic unit in the area, the Honduras–Nicaragua–NE Costa Rica metapopulation.

The Xcaret Ecoarchaeological Park is a private institution whose income comes from tourism, and that has been conducting ex situ reproduction of the species for the past 30 years [14]. It is registered as an UMA (Management Environment Unit) in the Wildlife Federal office in Mexico (Dirección General de Vida Silvestre, DGVS), under permit INE/CITES/DFYFS-ZOO-P-0011-99-Q.ROO. This permit does not allow commercial use of the macaws regarding direct sales, since this activity was banned in 2008, but allows the Park to manage its captive population for exhibition/education, and conservation purposes. Within this restriction, the Park has built many facilities for the macaws depending on the use of the specimens; some facilities are only used for night housing the macaws that are imprinted with humans and fly and return to the aviary and are not of reproductive age. Other facilities are used for macaws that are used for exhibition in close encounters with the public, and other facilities are for macaws that are reproducing.

Breeding at Xcaret is managed in aviaries of different sizes. At the beginning of the breeding program, pairs were artificially made by placing two adults together in the same cage, but later, bigger cages were built to include several adults of both sexes to let them choose their mates by themselves. Once the pair is formed, they are put in a cage of mid-size with a nesting box. Raising is performed both by hand and by their parents until they are three months old. There are climate-controlled facilities to raise nestlings by hand and more open facilities to feed youngsters until they feed themselves. All the macaws born in the population are banded with a closed marked steel ring according to the permit at 1–1.5 months old because, later, it would not be possible to insert the ring anymore. The ring has a unique key number and the Xcaret name. The reproductive output is reported annually to the DGVS with the list of rings placed with the specimens.

In a study of ex situ conservation genetics assessment, Witzenberger and Hochkirch [3] recommended that the number of founders must be a minimum of 15 and the population size at least 100 in order to minimize inbreeding problems in ex situ programs. The macaw population started with 29 pairs obtained from other captive populations in Mexico and from seized illegal specimens given to Xcaret by PROFEPA (Procuraduría Federal de Protección al Ambiente) for an approximate total of 60 individuals of unknown kinship. Breeding in captivity started in 1994 with six nestlings. The captive population grew to the goal of 100 breeding pairs, and more. The maximum number of macaws in the population was 946 (2012). As a result of different donations for reintroduction programs, the total number today is 596. Given these numbers, the captive population is expected to meet the criteria to avoid inbreeding.

Over the course of the first 20 years of management, decisions were made to maximize reproductive success by avoiding reproductive pairing of close relatives, using the record keeping system. Such decisions showed improvement regarding reproductive output and population growth. Although records were not kept at the beginning of the Park's operation, breeding data were recorded starting in 2007.

The actual breeding capacity now is 200 newborns per year but, due to space limitations, many nests are kept closed. All macaws born in captivity are issued a birth certificate with their parents' IDs and ring numbers. Available pedigree information has been used to avoid inbreeding.

Evaluation of the maintenance or loss of genetic diversity in the current ex situ population is very important, since this population is the foundation for reintroduction programs in Mexico (Palenque 2013–2018, Los Tuxtlas 2014–2021). This information could help optimize strategies to select individuals for reintroduction by maximizing genetic variation, thus avoiding bottlenecks and negative effects of inbreeding in the new populations to be established.

The objectives of the present study were to estimate the genetic diversity of the founders and breeding pairs of the Xcaret population in order to contribute to the conservation of this subspecies and to compile previous genetic studies to compare them with this captive population. To achieve this objective, we genotyped the founding individuals and breeding pairs with six microsatellite loci to therefore compare the diversity of both groups (founders and breeding pairs).

2. Materials and Methods

2.1. Sample Collection

Blood samples were obtained from 49 macaws identified as founders that are still alive, though they are no longer breeders. We also sampled the 166 scarlet macaws comprising the 83 breeding pairs of 2015, descendants of these founders. A drop of blood was collected from each individual and placed on labeled FTA cards (Whatman^{TM}, Florham, NJ, USA). All samples were collected at the Xcaret aviary and then sent to the laboratory for analysis and storage at room temperature. All samples were confirmed as *Ara macao cyanoptera* based on mitochondrial DNA (unpublished data).

2.2. DNA Extraction

DNA was extracted from blood samples using the proteinase K digestion technique with the kit and DNeasy Blood & Tissue® (Qiagen Valencia, Santa Clarita, CA, USA). Two snips of blood were used for lysis, 60 µL of PBS 1X, and 7 µL of proteinase K (1 mg/mL), adjusted at 220 µL of lysis buffer.

The extracted DNA was then quantified in a nucleic acid spectrophotometer (Nanodrop® Thermo, Wilmington, DE, USA) and visualized on 2% agarose gel. All samples had a final DNA concentration between 20 and 58 ng/µL at a purity of 1.5–1.9 (260/280 absorbance). Such quantity and purity are suitable to amplify microsatellite loci [15,16].

2.3. Amplification and Genotyping

Primers for six microsatellite loci designed for different species of *Ara* [17] were previously standardized in our laboratory (Table 1) and identified as variable and informative for *Ara macao cyanoptera*. Amplification of the loci was performed using the forward primer labeled with a fluorescent dye (VIC, FAM, PET, and NED, (Applied Biosystems® Foster City, CA, USA). PCR amplification was carried out with the Platinum® Taq DNA Polymerase kit (ThermoFisher, Waltham, MA, USA) with 1X Buffer, 2 mM MgCl$_2$, 0.8 µM of each primer, 0.2 µM dNTPs, 0.25 U taq polymerase, 1–2 µL of DNA (5–25 ng/µL), and finally bidistilled water to adjust a reaction volume of 12.5 µL. The reaction was performed with an initial denaturation of 95 °C for 30 s followed by 14 touchdown cycles and an annealing temperature of 60 °C for 30 s (with 0.5 °C per cycle decreases to 51 °C), 1 min at 72 °C, 30 cycles of 95 °C for 30 s, 51 °C for 30 s, and 72° C for 1 min, and a final extension of 7 min at 72 °C.

Table 1. Estimates of population genetic parameters for founders and breeding pairs from the Xcaret scarlet macaw population, including number of individuals (N), number of alleles (Na), effective number of alleles (Ne), observed heterozygosity (Ho), expected heterozygosity (He), private alleles (P), Nei's gene diversity (D), allelic richness (A_R), and inbreeding coefficient (F). * Significant ($p < 0.05$). ** Significant ($p < 0.05$) after correction using Bonferroni procedure.

Founders

Loci	N	Range	Na	Ne	Ho	He	P	D	A_R	F	HWE
AgGT17	48	115–137	10	5.592	0.875	0.821	-	0.832	9.997	−0.066	*/-
AgGT19	48	181–189	4	2.386	0.458	0.581	-	0.595	3.939	0.211	-
AgGT21	48	169–189	8	5.224	0.729	0.809	-	0.819	7.939	0.098	-
AgGT42	45	243–271	12	6.784	0.956	0.853	1	0.863	12	−0.121	-
UnaCT21	48	166–172	4	3.022	0.646	0.669	-	0.676	4	0.035	-
UnaCT74	48	150–168	7	2.808	0.625	0.644	2	0.661	6.936	0.029	*/**
Mean			7.5	4.303	0.717	0.729			7.466	0.031	
(SE)			1.31	0.735	0.074	0.046			1.316	0.048	

Breeding Pairs

Loci	N	Range	Na	Ne	Ho	He	P	D	A_R	F	HWE
AgGT17	159	115–137	12	6.11	0.881	0.836	2	0.839	10.508	−0.053	*/-
AgGT19	166	181–189	4	2.16	0.614	0.536	-	0.537	3.605	−0.147	*/-
AgGT21	155	169–191	9	4.87	0.858	0.795	1	0.797	8.093	−0.08	*/-
AgGT42	165	243–275	14	6.87	0.873	0.854	3	0.857	11.369	−0.022	*/**
UnaCT21	165	150–174	6	2.98	0.715	0.664	2	0.666	5.679	−0.077	*/**
UnaCT74	155	152–168	5	2.53	0.639	0.604	-	0.606	4.995	−0.058	-
Mean			8.3	4.250	0.763	0.715			7.375	−0.073	
(SE)			1.6	0.810	0.050	0.054			1.279	0.017	

The DNA amplicons obtained were sent to the Biodiversity and Health Genomic Sequencing Laboratory of the Institute of Biology of the Universidad Nacional Autónoma de México for genotyping. Each sample was prepared with 500 LIZ® Size Standard (Applied Biosystems) according to the manufacturer, and 1 μL of DNA amplicon per individual per loci, and then subsequently analyzed on an Applied Biosystems™ 3500xL laser sequencer (Life Technologies™, Carlsbad, CA, USA). Allele scoring was performed with GeneMapper v. 4.1 (Applied Biosystems™). All the fragments were analyzed twice to confirm the obtained results. Tandem v. 1.09 [18] was used to correct mobility problems and possible artifacts, which also confirmed the allelic assignment according to the repetition units of each locus. To detect null alleles, large allele dropout, and scoring errors, all loci were analyzed with MICRO-CHECKER v2.2.3 [19]. In GENEPOP Web [20], deviations from the Hardy–Weinberg equilibrium (HWE) caused by an excess or deficit of heterozygotes were analyzed using Fisher's exact test [21]. In GENALEX v. 6.5 [22], the linkage disequilibrium of all loci was assessed using the exact probabilities test. We estimated the significance level values of HWE and linkage disequilibrium by applying a Bonferroni correction ($\alpha = 0.01$). The following diversity indices were also obtained: average number of alleles per locus (Na), effective number of alleles (Ne), observed (Ho) and expected (He) heterozygosity, number of private alleles (P), and Nei's genetic diversity (D). Allelic richness (A_R), standardized with the smallest sample size (46), was obtained in FSTAT v. 2.9.4 [23]. Each index was calculated per locus and per scarlet macaw group studied (founders and breeding pairs).

2.4. Population Structure

The population structure was examined through a principal coordinates analysis (PCoA) based on Nei's genetic distances matrix between individuals using GENALEX ver 6.5. In addition, population structure was examined using STRUCTURE v.2.3 software [24] to perform a Bayesian clustering method in order to infer the number of clusters and structure. An admixture model with the LOCPRIOR option was used. The number of tested populations (K) ranged from 1 to 10 using 20 independent Markov Chain Monte Carlo (MCMC), by sampling 200,000 iterations, and a 200,000 burn-in period. The most likely number of clusters was inferred using STRUCTURE HARVESTER [25]. To infer the optimal K-value, both ln Pr (K) and the ΔK statistic [26] were calculated. To further search for genetic clusters without assuming an underlying population genetic model such as HW equilibrium or linkage disequilibrium, we applied the discriminant analysis of principal components (DAPC, [27]), performed with the package ADEGENET v 2.2.1 [28] of RStudio v 1.4.1717 (©2009–2021 RStudio, PBC) in R v. 4.1.1 [29]. We explored the data with K = 10, K = 20 and K = 30. The lowest value of the Bayesian information criterion (BIC) was used as the number of clusters that best reflect the population structure of the data.

To estimate the degree of kinship (r) between individuals and populations (founders and breeding pairs), we evaluated different "relatedness r" estimators available in the GENALEX software. To compare the informativeness and the power of relatedness detection of available estimators, we used the reciprocal of the mean squared deviations (RMSD) of estimators provided in the KinInfor program v.2 [30]. Our objectives were to detect full sibs ($\Delta 1$ = 0.5, $\Delta 2$ = 0.5), paternal halfsibs ($\Delta 1$ = 1, $\Delta 2$ = 0), maternal half sibs ($\Delta 1$ = 0.5, $\Delta 2$ = 0), and unrelated ($\Delta 1$ = 0, $\Delta 2$ = 0) individuals. We ran simulated pairs of genotypes and set the confidence level at 0.05. For each estimator, the mean and standard deviation were calculated. The COANCESTRY program [31] was used to compute the R values for each pair of individuals (dyads) and evaluate the statistical errors associated with these estimates using bootstrapping (1000 replicates).

3. Results

Genotypes were obtained for six microsatellite loci from a total of 49 founders and 166 breeding individuals of Mesoamerican scarlet macaws. No evidence of null alleles or allelic dropout was found. Linkage disequilibrium (LD) was observed between 11 of 30 loci comparisons and was observed only in the breeding population (Supplementary Materials Table S1). Deviations from the Hardy–Weinberg equilibrium were detected only in AgGT17 and UnaCT74 in the founder population, but in all loci, except UnaCT74 in the breeding pairs group, after Bonferroni correction (Table 1 and Table S2). Based on the overall test, the founder population showed a deficit of heterozygotes (p = 0.0250), while the breeding pairs showed an excess of heterozygotes (p = 0.0164).

In terms of genetic diversity, the six loci were polymorphic for both populations. The number of private alleles was greater in the breeding population. (8 vs. 3). The diversity (D) and allelic richness (A_R) differed very slightly (Table 1). The comparison of allelic richness was not significantly different between founders and breeding populations (A_R = 7.466 vs. A_R = 7.375, p = 0.96), neither in observed heterozygosity (Ho = 0.717 vs. 0.763, p = 0.8325) nor gene diversity (D = 0.731 vs. D = 0.717, p = 0.496).

The pattern of genetic structure determined by PCoA displays the overlap between founder and breeding pair groups (Figure 1, Supplementary Materials Figure S1). The Bayesian analysis of population structure revealed that the maximum value of ln P (K) obtained was K = 5 (Figure 2a), which is concordant with the maximum Delta K (Figure 2b) (Supplementary Materials Table S3). The individuals of the founder group are formed by different proportions of the five genetic lineages observed (Figure 2c); however, there is a higher prevalence of lineages 1 (red) and 2 (green), both in the founders and in the breeders (Figure S1).

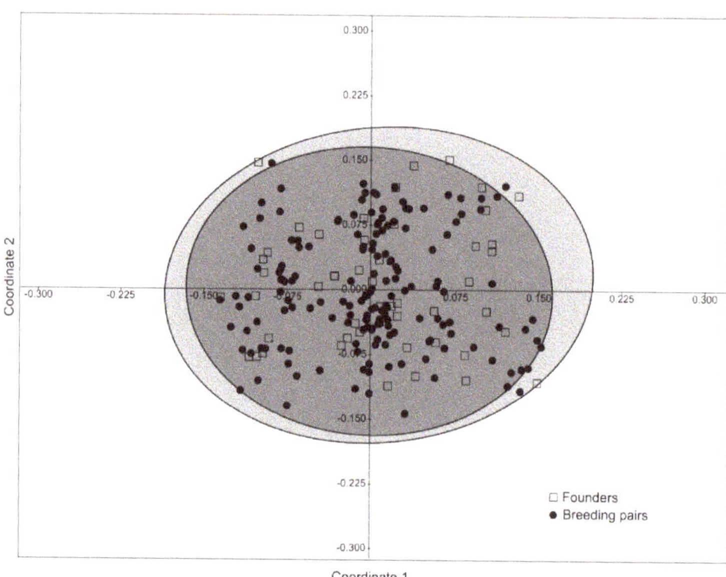

Figure 1. PCoA analysis of 215 individuals of founders (*n* = 49) and breeding pairs (*n* = 116) in the Xcaret captive population of Mesoamerican scarlet macaw (*Ara macao cyanoptera*) based on genotyping results of six microsatellites markers.

In contrast to STRUCTURE results, eight gene clusters were obtained with DAPC (Figure 3) for K = 10, K = 20, and K = 30 analysis (Supplementary Materials, Figure S1). However, the separation of the clusters is not entirely clear (Supplementary Materials, Figure S1); clusters 1, 2, 4, 5, and 7 overlap extensively, but 3, 8, and specially 6 are more clearly defined.

Comparisons among the informativeness of different relatedness estimators yielded Ritland [32] as the best estimator (Tables S4–S6). Ranking of the loci according to informativeness was: AgGT42, AgGT17, AgGT21, UnaCT21, UnaCT74, and AgGT19. Mean and variance of the different relatedness estimators are provided in Supplementary Materials Tables S4 and S5. The averaged relatedness values are no greater than −0.004 in founders and −0.002 in breeding pairs (Figure 4). The relatedness values in both groups are within the expected estimated confidence intervals (Figure 4). Mean relatedness was lower in the founders but slightly increased in the breeding pairs, perhaps due to the presence of siblings in the group (although R is close to zero, indicating most individuals are unrelated).

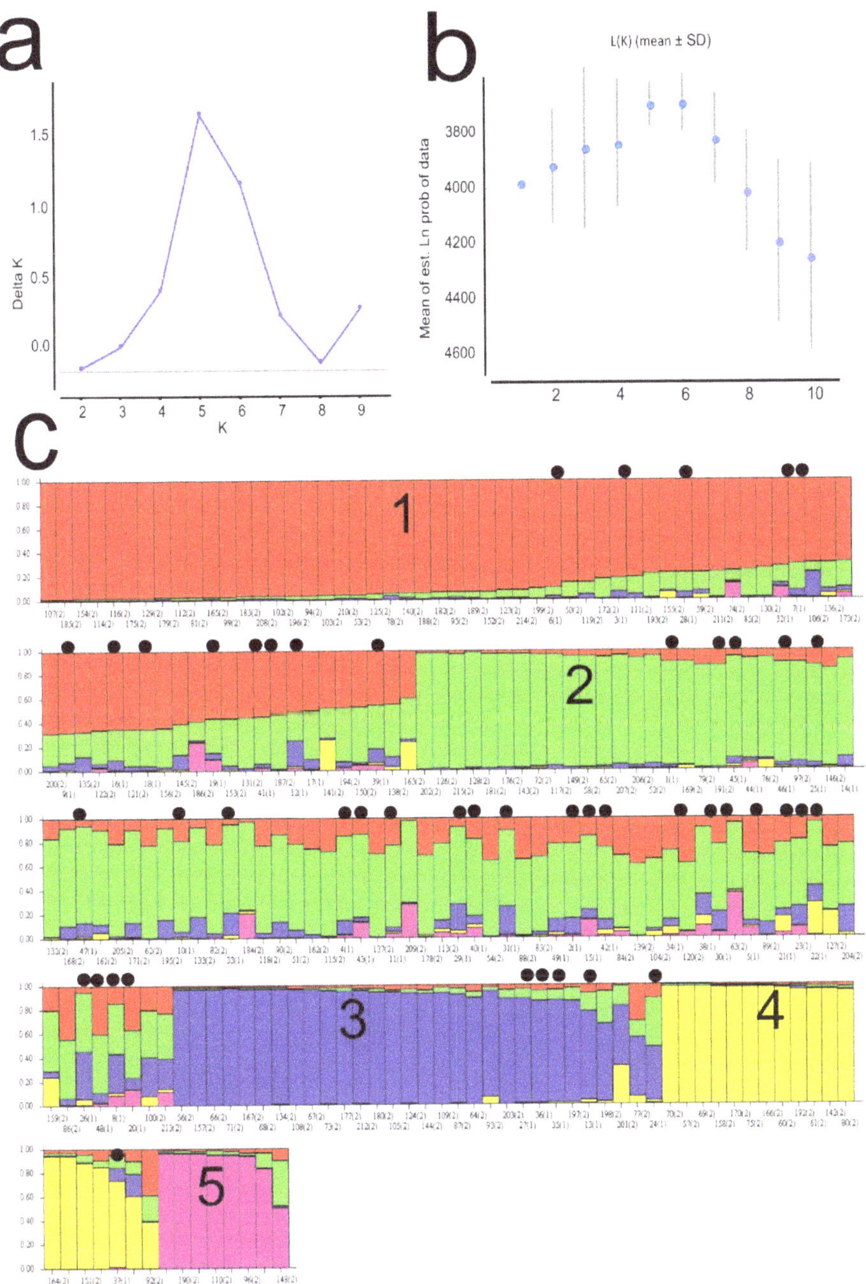

Figure 2. (**a**) Mean ln P (K) graph, (**b**) Evannos ΔK graph, (**c**) Q-membership proportion of K = 5 genetic clusters of founders (black dots), and breeding individuals of the Xcaret Mesoamerican scarlet macaw (*Ara macao cyanoptera*) based in six microsatellites markers. The length of the colored segment indicates the proportion of the individual's composition in specific clusters showing admixture in the population.

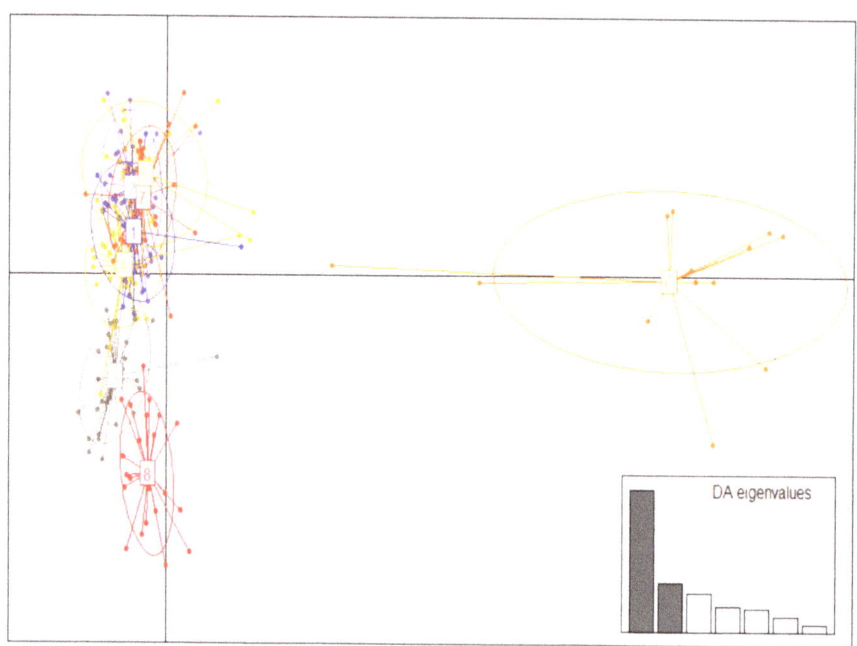

Figure 3. Gene clusters found with the DAPC analysis showing some structure in the captive population of scarlet macaws of Xcaret.

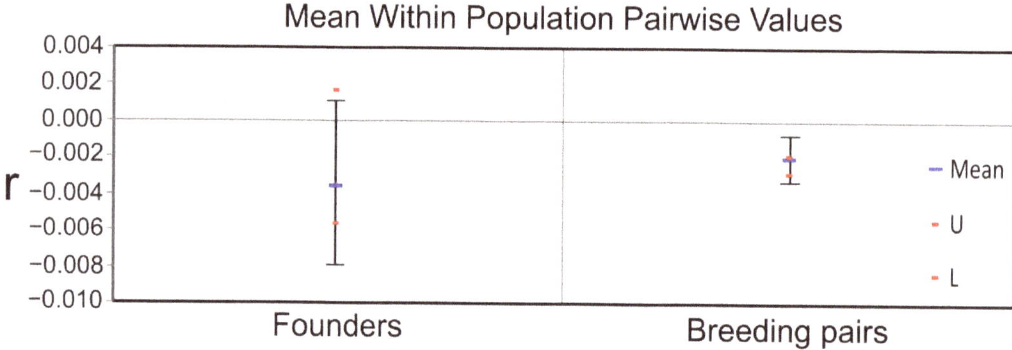

Figure 4. Ritland's relatedness (r) index of the Xcaret Mesoamerican scarlet macaw (*Ara macao cyanoptera*) in founders and in breeding pairs groups.

4. Discussion

The estimates of the population genetic parameters for founders and breeding pairs from the Xcaret scarlet macaw population were very similar. The slightly larger number of alleles in breeding pairs probably reflects the fact that we were not able to obtain samples from all of the founder individuals, since some had died before our study was carried out. Reproduction in the species extends until they are 35 years old, with a longevity record of up to 64–65 years [33]; thus, the number of generations in the population is small. Reproduction begins when individuals are approximately four years old (Xcaret, pers. com.) and generations overlap; this feature has helped preserve the original genetic diversity of the population up to the present time. Populations in captivity have often been observed to depart from the HW equilibrium [34], and in this ex situ population, significant departures were noticeable in the breeding pairs in comparison with the founders. These results are perhaps accounted for by nonrandom mating management in captivity in this population.

The genetic diversity of some wild populations of *A. macao* have already been studied using the same microsatellites. In addition to trying microsatellites for different species of *Ara* and *Amazona*, Gebhardt and Waits [17] tested *Ara m. macao* from Peru. From a wild population of *A. m. macao*, scarlet macaws were sampled in Pará (Brazil) with 10 microsatellites [35]. From a population of the Mayan Forest in Guatemala, 11 microsatellites were tested [36]. Additionally, two populations from Costa Rica (*A. m. macao*) were studied [37] using six microsatellites (Table 2).

Table 2. Population origin, sample size (N), number of alleles (A), and expected (He) and observed (Ho) heterozygosities of microsatellite loci in the two subspecies of scarlet macaws studied to date.

Species/Population	Origin	N	Na	He	Ho	Reference
A. macao cyanoptera (Mesoamerican scarlet macaw)						
Founders (Xcaret)	Captivity	49	7.5	0.715	0.729	This study
Breeding individuals	Captivity	166	8.3	0.763	0.715	This study
La Selva Maya, Guatemala	Wild	37	7.1	0.696	0.713	[33]
A. m. macao (scarlet macaw)						
Costa Rica (CP)	Wild	41	6.7	0.63	0.61	[34]
Costa Rica (SP)	Wild	55	8.1	0.68	0.65	[34]
Perú	Wild	25	10.3	0.833	0.84	[16]
Brazil	Wild	28	11.8	0.846	0.842	[32]

Moderate levels of genetic diversity were found in the Costa Rican wild populations, similar to those of Guatemala, with indications of imbalance possibly due to genetic erosion caused by anthropogenic factors that are demographically affecting them. More stable populations in the Amazon (Peru and Brazil) have more genetic variability, leading to the interpretation that the scarlet macaw is a generalist species and until recently was widely distributed, exhibiting high genetic diversity in relation to other species of more specialized macaws with less diversity [35,38].

Comparing these data on wild populations with those of the Xcaret founders and breeding pairs, heterozygosity and number of allele values are very similar, but wild populations in Guatemala (*cyanoptera*) and Costa Rica (subspecies *macao*) have slightly lower values, whereas those of *macao* populations in Brazil and Peru, where populations are still large, are slightly higher. This similarity in the genetic variability values of the founding and reproductive populations of Xcaret with those of wild populations indicates that the ex situ population maintains levels of genetic diversity comparable to wild populations of the subspecies. However, these comparisons are not tested statistically.

Relatively large population sizes [39], high dispersal [40], and longevity [33,41] may have helped buffer scarlet macaw populations against genetic erosion. In fact, the Xcaret population is now two times larger than the Mayan Forest population, as estimated by Boyd and McNab [11].

Between the captive groups of founders and breeding pairs, there is a lack of population structure, which was evident with the PCoA and the Bayesian analysis, in which admixture is evident in the genetic composition of the individuals; hence, there is no genetic differentiation between founding and current breeding pairs in the captive population. However, the statistics obtained with STRUCTURE indicate that there are five genetic lineages that may be the result of a previous genetic structure in the wild populations where the founders came from. A similar population structure was found with the DAPC analysis, a method that emphasizes the differences between clusters. Of the eight clusters identified, five were not distinguishable in the graphic but three others were, perhaps reflecting a historical population structure from the original wild populations. The genetic diversity of the ex situ populations is determined by the gene pool of the founders and their reproductive success [42]. Using mitochondrial DNA sequence data, Schmidt et al. [13] showed there was geographic clustering of haplotypes that might suggest genetic structuring of the wild populations of *A. m. cyanoptera* whom the founders of the Xcaret population came from.

Ex situ conservation programs strive to maintain genetic diversity that is comparable to a wild population by capturing a sufficient number of founders and managing matings to select individuals with underrepresented genes [43]. The objective of any reintroduction program is to provide enough genetic diversity to circumvent negative effects of natural selection within the new population, assuming that the reintroduced population can grow rapidly. If a source population has low genetic diversity, the reintroduced population that is derived from it will have similarly low diversity [44]. Our data show that the captive macaw population at Xcaret has captured and maintained similar levels of genetic variability to wild macaw populations of the Selva Maya. An important next step would be to compare the breeding population to the wider natural population. This analysis might reveal that there are unsampled populations in the wild such as The Chimalapas (Oaxaca, or Selva Maya W in [36]); as such, the captive breeding program may be improved by bringing in further birds from the wild to increase the genetic representation of the captive breeding population. Another possibility would be that representatives of this western population survive in captivity in some other aviaries and could be traced.

With limited pedigree information, at least initially, the reproductive management of the captive reproductive population has followed the strategy of reducing kinship to favor reproductive success and, unintentionally, conserving maximum genetic diversity within the population [45]. Hence, with the reproductive management carried out in Xcaret, it has been possible to maintain the gene pool of the founders in the reproductive population. Therefore, it is not surprising that an underlying structure was also found with microsatellites with five clusters that are well marked but intertwining.

The inbreeding coefficient (F) was low (<0.03) for both groups (founders and breeding pairs), which indicates that matings between closely related individuals is infrequent in both populations; however, the pairwise relatedness value in the breeding population showed a slight increase. Although this increase was not significant, it could indicate that ongoing careful management of breeding pairs will be required to avoid inbreeding. Although the breeding pairs group have generally avoided inbreeding, and the population has maintained levels of genetic variability comparable to those found in wild populations, it is recommended to identify individuals with a high degree of kinship in order to prevent breeding attempts by closely related individuals. The kinship information obtained in this study with the breeding pairs, complemented by pedigree information for the descendants of these pairs, should be used for further growth of the captive colony, as well as for the selection of group compositions for reintroduction projects with the aim of providing maximum genetic variability in the new populations, avoiding bottlenecks and promoting the success of reintroductions from a genetic viewpoint.

In order to secure the long-term persistence of reintroduced populations, it is also important that ex situ breeding programs endeavor to minimize time in captivity. If programs exceed the limit of 10–15 generations, relaxed selection could incur genetic costs [46]. With these guidelines, the Xcaret and its allies' opening of reintroduction projects of the subspecies in two separate sites in Mexico (Palenque, Chiapas and Los Tuxtlas, Veracruz) is timely.

Supplementary Materials: The following supporting information can be downloaded at: https://www.mdpi.com/article/10.3390/d14010054/s1, Figure S1: Genetic clusters obtained from the STRUCTURE analysis; Table S1: Genotypic linkage disequilibrium; Table S2: Summary ofor Hardy-Weinberg Equilibrium; Table S3: Evanno summary; Table S4: RMSD in KinInfor; Table S5: Relatedness estimates in Coancestry; Table S6: Ritland's relatedness estimator; Table S7 Table of alleles.

Author Contributions: P.E.-P. designed the study, coordinated the work, and participated in the analysis and interpretation of results as well as the writing of the manuscript. N.M.-F. performed the laboratory assays, analyzed the data and participated in the interpretation of results and writing of the manuscript. P.R.-E. carried out part of the laboratory assays and participated in the interpretation of results and writing of the manuscript. G.L.-M., K.S.-G. and R.R.-F. maintained the reproductive population, helped collect samples, and participated in the interpretation of results and writing of the manuscript. All authors have read and agreed to the published version of the manuscript.

Funding: Part of the funding for this work was granted by the Priority Species Program of the National Commission of Natural Protected Areas (SEMARNAT, grant PROCER/DGOR/26/2014).

Institutional Review Board Statement: Not applicable.

Data Availability Statement: The data presented in this study are available in Supplementary Table S7: Table of alleles.

Acknowledgments: We wish to thank Bosque Antiguo AC for its support to obtain and handle the resources of the project, as well as Jonathan Morales-Contreras and Pilar Luna for their committed assistance in the lab and the aviary. We also thank Laura Márquez-Valdelamar for her assistance with the genotyping of the data. We likewise thank Guadalupe Amancio Rosas for her support in performing part of the analysis. We also thank comments from three reviewers that improved the original manuscript substantially.

Conflicts of Interest: There is no conflict of interest from the authors.

References

1. IUCN/SSC Guidelines for reintroductions and other conservation translocations. In *Gland Switz Camb UK IUCNSSC Re-Introd Spec Group*; IUCN: Gland, Switzerland; Cambridge, UK, 2013; p. 57.
2. Boyce, W.M.; Weisenberger, E.M.; Penedo, M.C.T.; Johnson, C.K. Wildlife translocation: The conservation implications of pathogen exposure and genetic heterozygosity. *BMC Ecol.* **2011**, *11*, 5. [CrossRef]
3. Witzenberger, K.A.; Hochkirch, A. Ex situ conservation genetics: A review of molecular studies on the genetic consequences of captive breeding programmes for endangered animal species. *Biodivers. Conserv.* **2011**, *20*, 1843–1861. [CrossRef]
4. Spielman, D.; Brook, B.; Frankham, R. Most species are not driven to extinction before genetic factors impact them. *Proc. Natl. Acad. Sci. USA* **2004**, *101*, 15261–15264. [CrossRef]
5. SEMARNAT (Secretaría de Medio Ambiente y Recursos Naturales). Norma Oficial Mexicana NOM-059-SEMARNAT-2010, Protección Ambiental-Especies Nativas de México de Flora y Fauna Silvestres-Categorías de Riesgo y Especificaciones para su Inclusión, Exclusión o Cambio-Lista de Especies en Riesgo. DOF del 30 de Diciembre de 2010. Available online: https://www.dof.gob.mx/nota_detalle.php?codigo=5578808&fecha=14/11/2019 (accessed on 10 September 2021).
6. Wiedenfeld, D.A. A new subspecies of Scarlet Macaw and its status and conservation. Una nueva subespecie de la lapa roja y su estado de conservación. *Ornitol. Neotrop.* **1994**, *5*, 99–104.
7. SEMARNAT (Secretaría de Medio Ambiente y Recursos Naturales). Programa de Acción para la Conservación de la Especie Guacamaya Roja *(Ara macao cyanoptera)*. Comisión Nacional de Áreas Naturales, Protegidas, Secretaría del Medio Ambiente y Recursos Naturales. 2009. Available online: https://www.conanp.gob.mx/pdf_especies/Pace_Guacamaya_Roja.pdf (accessed on 10 September 2021).
8. CITES. Check-List of CITES Species. 2013. Available online: http://checklist.cites.org/# (accessed on 10 September 2021).
9. Snyder, N.; McGowan, P.; Gilardi, J.; Grajal, A. (Eds.) *Parrots. Status Survey and Conservation Action Plan 2000–2004*; IUCN: Gland, Switzerland; Cambridge, UK, 2000; p. 180.
10. Berkunsky, I.; Quillfeldt, P.; Brightsmith, D.; Abbud, M.; Aguilar, J.; Alemán-Zelaya, U.; Aramburú, R.; Arias, A.A.; McNab, R.B.; Balsby, T.J.S.; et al. Current threats faced by Neotropical parrot populations. *Biol. Conserv.* **2017**, *214*, 278–287. [CrossRef]
11. Boyd, J.D.; McNab, R.B. The Scarlet Macaw in Guatemala and El Salvador: 2008 Status and Future Possibilities. In *Findings and Recommendations from a Species Recovery Workshop 9–15 March 2008*; Guatemala City and Flores, Petén, Guatemala; Wildlife Conservation Society: Bronx, NY, USA, 2008.
12. García-Feria, L.M. Un Enfoque Filogeográfico Para la Conservación de Poblaciones de Ara Macao Cyanoptera. Ph.D. Thesis, Instituto de Ecología, AC, Xalapa, Mexico, 2009.
13. Schmidt, K.L.; Aardema, M.L.; Amato, G. Genetic analysis reveals strong phylogeographical divergences within the Scarlet Macaw Ara macao. *Ibis* **2020**, *162*, 735–748. [CrossRef]
14. Raigoza-Figueras, R. Scarlet macaw *Ara macao cyanoptera* conservation programme in Mexico. *Int. Zoo Yearb.* **2014**, *48*, 48–60. [CrossRef]
15. Paetkau, D.; Calvert, W.; Stirling, I.; Strobeck, C. Microsatellite analysis of population structure in Canadian polar bears. *Mol. Ecol.* **1995**, *4*, 347–354. [CrossRef] [PubMed]

16. Russello, M.; Calcagnotto, D.; DeSalle, R.; Amato, G. Characterization of microsatellite loci in the endangered St. Vincent Parrot, Amazona guildingii. *Mol. Ecol. Notes* **2001**, *1*, 162–164. [CrossRef]
17. Gebhardt, K.J.; Waits, L.P. Cross-species amplification and optimization of microsatellite markers for use in six Neotropical parrots. *Mol. Ecol. Resour.* **2008**, *8*, 835–839. [CrossRef] [PubMed]
18. Matschiner, M.; Salzburger, W. TANDEM: Integrating automated allele binning into genetics and genomics workflows. *Bioinformatics* **2009**, *25*, 1982–1983. [CrossRef] [PubMed]
19. VAN Oosterhout, C.; Hutchinson, W.F.; Wills, D.P.M.; Shipley, P. micro-checker: Software for identifying and correcting genotyping errors in microsatellite data. *Mol. Ecol. Notes* **2004**, *4*, 535–538. [CrossRef]
20. Raymond, M.; Rousset, F. GENEPOP (version 1.2): Population genetics software for exact tests and ecumenicism. *J. Hered.* **1995**, *86*, 248–249. [CrossRef]
21. Guo, S.W.; Thompson, E.A. Performing the exact test of Hardy-Weinberg proportion for multiples alleles. *Biometrics* **1992**, *48*, 361–372. [CrossRef]
22. Peakall, R.; Smouse, P.E. GenAlEx 6.5: Genetic analysis in Excel. Population genetic software for teaching and research—An update. *Bioinformatics* **2012**, *28*, 2537–2539. [CrossRef] [PubMed]
23. Goudet, J.F. FSTAT (version 1.2): A computer program to calculate F-statistics. *J. Hered.* **1995**, *86*, 485–486. [CrossRef]
24. Pritchard, J.K.; Stephens, M.; Donnelly, P. Inference of population structure using multilocus genotype data. *Genetics* **2000**, *155*, 945–959. [CrossRef] [PubMed]
25. Earl, D.A.; vonHoldt, B.M. Structure Harvester: A website and program for visualizing STRUCTURE output and implementing the Evanno method. *Conserv. Genet. Resour.* **2012**, *4*, 359–361. [CrossRef]
26. Evanno, G.; Regnaut, S.; Goudet, J. Detecting the number of clusters of individuals using the software structure: A simulation study. *Mol. Ecol.* **2005**, *14*, 2611–2620. [CrossRef]
27. Jombart, T.; Devillard, S.; Balloux, F. Discriminant analysis of principal components: A new method for the analysis of genetically structured populations. *BMC Genet.* **2010**, *11*, 94. [CrossRef]
28. Jombart, T. Adegenet: A R package for the multivariate analysis of genetic markers. *Bioinformatics* **2008**, *24*, 1403–1405. [CrossRef]
29. R Core Team. *R: A Language and Environment for Statistical Computing. R Foundation for Statistical Computing*; R Core Team: Vienna, Austria, 2020. Available online: https://www.R-project.org/ (accessed on 15 November 2021).
30. Wang, J. Informativeness of genetic markers for pairwise relationship and relatedness inference. *Theor. Popul. Biol.* **2006**, *70*, 300–321. [CrossRef] [PubMed]
31. Wang, J. Coancestry: A program for simulating, estimating and analysing relatedness and inbreeding coefficients. *Mol. Ecol. Resour.* **2010**, *11*, 141–145. [CrossRef]
32. Ritland, K. Estimators for pairwise relatedness and individual inbreeding coefficients. *Genet. Res.* **1996**, *67*, 175–185. [CrossRef]
33. Brouwer, K.; Jones, M.L.; King, C.E.; Schifter, H. Longevity records for Psittaciformes in captivity. *Int. Zoo Yearb.* **2000**, *37*, 299–316. [CrossRef]
34. Gonçalves, E.C.; Ferrari, S.F.; Bastos, H.B.; Wajntal, A.; Aleixo, A.; Schneider, M.P. Comparative genetic diversity of wild and captive populations of the bare-faced curassow (Crax fasciolata) based on cross-species microsatellite markers: Implications for conservation and management. *Biochem. Genet.* **2010**, *48*, 472–479. [CrossRef] [PubMed]
35. Presti, F.T.; Oliveira-Marques, A.R.; Caparroz, R.; Biondo, C.; Miyaki, C.Y. Comparative analysis of microsatellite variability in five macaw species (Psittaciformes, Psittacidae): Application for conservation. *Genet. Mol. Biol.* **2011**, *34*, 348–352. [CrossRef]
36. Schmidt, K. Spatial and Temporal Patterns of Genetic Variation in Scarlet Macaws (Ara Macao): Implications for Population Management in La Selva Maya. Ph.D. Thesis, Columbia University, New York, NY, USA, 2013.
37. Monge, O.; Schmidt, K.; Vaughan, C.; Gutiérrez-Espeleta, G. Genetic patterns and conservation of the Scarlet Macaw (Ara macao) in Costa Rica. *Conserv. Genet.* **2016**, *17*, 745–750. [CrossRef]
38. Caparroz, R.; Miyaki, C.Y.; Bampi, M.I.; Wajntal, A. Analysis of the genetic variability in a sample of the remaining group of Spix's Macaw (Cyanopsitta spixii, Psittaciformes: Aves) by DNA fingerprinting. *Biol. Conserv.* **2001**, *99*, 307–311. [CrossRef]
39. Dear, F.; Vaughan, C.; Polanco, A.M. Current Status and Conservation of the Scarlet Macaw (Ara macao) in the Osa Conservation Area (ACOSA), Costa Rica. *UNED Res. J.* **2010**, *2*, 7–21. [CrossRef]
40. Myers, M.C.; Vaughan, C. Movement and behavior of scarlet macaws (Ara macao) during the post-fledging dependence period: Implications for in situ versus ex situ management. *Biol. Conserv.* **2004**, *118*, 411–420. [CrossRef]
41. Bourke, B.P.; Frantz, A.C.; Lavers, C.P.; Davison, A.; Dawson, D.A.; Burke, T. Genetic signatures of population change in the British golden eagle (Aquila chrysaetos). *Conserv. Genet.* **2010**, *11*, 1837–1846. [CrossRef]
42. Witzenberger, K.A.; Hochkirch, A. Evaluating ex situ conservation projects: Genetic structure of the captive population of the Arabian sand cat. *Mamm. Biol.* **2013**, *78*, 379–382. [CrossRef]
43. Haig, S.M.; Ballou, J.D.; Derrickson, S.R. Management Options for Preserving Genetic Diversity: Reintroduction of Guam Rails to the Wild. *Conserv. Biol.* **1990**, *4*, 290–300. [CrossRef]
44. Jamieson, I.G. Founder Effects, Inbreeding, and Loss of Genetic Diversity in Four Avian Reintroduction Programs. *Conserv. Biol.* **2011**, *25*, 115–123. [CrossRef] [PubMed]
45. Saura, M.; Pérez-Figueroa, A.; Fernández, J.; Toro, M.A.; Caballero, A. Preserving Population Allele Frequencies in Ex Situ Conservation Programs. *Conserv. Biol.* **2008**, *22*, 1277–1287. [CrossRef]
46. Robert, A. Captive breeding genetics and reintroduction success. *Biol. Conserv.* **2009**, *142*, 2915–2922. [CrossRef]

Opinion

Reducing the Extinction Risk of Populations Threatened by Infectious Diseases

Gael L. Glassock [1], Catherine E. Grueber [1], Katherine Belov [1] and Carolyn J. Hogg [1,2,*]

[1] School of Life & Environmental Sciences, The University of Sydney, Sydney, NSW 2006, Australia; ggla5270@uni.sydney.edu.au (G.L.G.); catherine.grueber@sydney.edu.au (C.E.G.); kathy.belov@sydney.edu.au (K.B.)

[2] San Diego Zoo Global, San Diego, CA 92101, USA

* Correspondence: carolyn.hogg@sydney.edu.au; Tel.: +61-2-8627-4896

Abstract: Extinction risk is increasing for a range of species due to a variety of threats, including disease. Emerging infectious diseases can cause severe declines in wild animal populations, increasing population fragmentation and reducing gene flow. Small, isolated, host populations may lose adaptive potential and become more susceptible to extinction due to other threats. Management of the genetic consequences of disease-induced population decline is often necessary. Whilst disease threats need to be addressed, they can be difficult to mitigate. Actions implemented to conserve the Tasmanian devil (*Sarcophilus harrisii*), which has suffered decline to the deadly devil facial tumour disease (DFTD), exemplify how genetic management can be used to reduce extinction risk in populations threatened by disease. Supplementation is an emerging conservation technique that may benefit populations threatened by disease by enabling gene flow and conserving their adaptive potential through genetic restoration. Other candidate species may benefit from genetic management via supplementation but concerns regarding outbreeding depression may prevent widespread incorporation of this technique into wildlife disease management. However, existing knowledge can be used to identify populations that would benefit from supplementation where risk of outbreeding depression is low. For populations threatened by disease and, in situations where disease eradication is not an option, wildlife managers should consider genetic management to buffer the host species against inbreeding and loss of genetic diversity.

Keywords: genetic rescue; genetic restoration; supplementation; disease; genetic diversity; inbreeding; conservation

Citation: Glassock, G.L.; Grueber, C.E.; Belov, K.; Hogg, C.J. Reducing the Extinction Risk of Populations Threatened by Infectious Diseases. *Diversity* 2021, 13, 63. https://doi.org/10.3390/d13020063

Academic Editor: Michael Wink
Received: 30 November 2020
Accepted: 2 February 2021
Published: 4 February 2021

Publisher's Note: MDPI stays neutral with regard to jurisdictional claims in published maps and institutional affiliations.

1. Introduction

Species extinction is a serious and pressing environmental challenge [1]. Loss of biodiversity disrupts ecosystem functioning, damages ecosystem services, and impacts human wellbeing [2,3]. Along with habitat fragmentation, small population sizes, invasive species, and climate change, emerging infectious diseases (hereafter diseases) contribute to species' decline [4,5]. Whilst there are no known instances of species extinction solely due to disease, black-footed ferrets (*Mustela nigripes*), Polynesian tree snails (*Partula nodosa*), Hawaiian honeycreepers (*Carduelinae*), African lions (*Panthera leo*), and American chestnuts (*Castanea dentata*) are some of the species that have experienced population crashes due to disease [6,7]. By decreasing the size of host populations, disease outbreaks can reduce their fitness and adaptability, predisposing them to extinction from other threats [4,5,8]. The reduced viability of disease-affected populations occurs because small populations are susceptible to inbreeding and genetic drift, which can result in increased homozygosity, fixation of deleterious alleles, and loss of genetic diversity [9–12]. Fragmented habitats can further exacerbate the impact of disease by reducing the movement of individuals between populations [13]. Whilst disease is a known threatening process that contributes to the endangerment and extinction of species [4,5], it can be difficult to mitigate.

Current efforts to reduce the impact of diseases are generally focused on direct interventions, such as vaccines and/or host treatments [14]. Human intervention using these management practices can be successful in preserving populations affected by disease [5,14]. For example, the Arctic fox (*Vulpes lagopus*) variant of rabies was eradicated in eastern Ontario, Canada, following delivery of an oral rabies vaccine [15]. In contrast, some diseases are difficult to eradicate [16]. It is currently not possible to eliminate the devastating amphibian pathogen, chytrid fungus (*Batrachochytrium dendrobatidis*), due to environmental reservoirs and multiple host species sustaining the pathogen within the environment [17]. As a result, current management practices may limit the impact of a disease, but not eradicate it. Therefore, what should happen when it is not possible to directly reduce the prevalence and impact of a disease? Host populations may recover if a treatment is developed or pathogen–host coevolution occurs [18,19]. However, until some degree of pathogen immunity emerges, whether naturally or via human intervention such as a vaccine, many populations that are suffering significant disease outbreaks remain vulnerable and at low densities. Repopulating an area with disease-free individuals following disease-induced local population extinction has been successful in some instances [20]. However, this method is often unrealistic because it is dependent on the existence of a healthy source population. We suggest that, when pathogen eradication is not a viable management strategy, an alternative to preventing population extinctions is to genetically manage host populations until the disease can be more effectively controlled.

2. The Tasmanian Devil

2.1. Devil Facial Tumour Disease (DFTD)

Most infectious diseases are caused by pathogens, an organism capable of being transmitted from host-to-host where it causes disease. However, disease threats come in many forms. Tasmanian devils (*Sarcophilus harrisii*) are an example of a species that has suffered widespread decline following the outbreak of a disease. Endemic to the island state of Tasmania, devils are the world's largest surviving carnivorous marsupial. They are threatened by devil facial tumour disease (DFTD), which is a transmissible cancer, where the cancer cell line is the infectious agent and is passed as an allograft between individuals [21]. DFTD is characterised by large, ulcerating facial tumours [22], and is nearly always fatal [23,24]. Since DFTD was first detected in the mid-1990s, devil numbers have declined by 80 percent in the wild [25]. Precocial breeding by one-year-old females appears to be maintaining devil populations, despite the presence of DFTD [25,26]. However, contemporary modelling has shown that many of these populations are susceptible to small population pressures, including inbreeding [27], making them less adaptable and more vulnerable to other threatening processes. Tasmanian devils are listed as endangered both internationally on the International Union for Conservation of Nature (IUCN) Red List [28], and nationally under the Environment Protection and Biodiversity Conservation Act 1999, and require management to mitigate the threat of DFTD and to maintain the species in the landscape. Although immunisation trails have successfully induced an anti-DFTD antibody response in disease-free devils [29], research into a vaccine against DFTD is ongoing. As such, it is currently not possible to directly alleviate the negative impact DFTD has on devil populations in the wild.

2.2. Genetic Managment

The Save the Tasmanian Devil Program (STDP) is the official government response to DFTD. Established in 2003, one of its aims is to maintain devils and their ecological function in the wild. In the absence of effective treatments or vaccines for DFTD, the decision was made to manage devil populations in the presence of disease, rather than the disease itself as per general disease management strategies [30]. The remnant devil populations are fragmented, with most showing low connectivity [27,31], low genetic diversity [32] and population decline [25]; therefore, they are considered a candidate for genetic management.

Translocations involve the human-mediated movement of individuals from one location to another, with the aim of supplying a population-level or species-level benefit [33]. Supplementation is a specific type of translocation where individuals are released into an existing population of conspecifics [33]. Supplementation offers a potential mechanism to artificially implement gene flow into disease populations and reverse the negative genetic changes that have occurred due to low population densities [34–38]. Introducing novel alleles via supplementation can improve allelic richness and increase genetic diversity, boosting the adaptability of the population and preventing deleterious alleles from becoming fixed [34,37,39–43]. Supplementation may also increase the number of individuals in the recipient population, buffering it against genetic drift and stochastic events [44,45]. Some have argued that the persistence of a few small, isolated populations in the absence of gene flow is evidence against the need for supplementing some populations with low genetic diversity [46,47]. However, these arguments are not supported by substantial empirical evidence and may be damaging to future conservation efforts [48]. Supplementation may be particularly beneficial for populations threatened by disease, because of gene flow into a host population, providing additional host alleles upon which selection can act [49].

The STDP established the Wild Devil Recovery Project in 2015 to investigate the use of supplementations to genetically manage declining wild devil populations [30]. Modelling has predicted that ongoing supplementation of DFTD-affected wild devil populations will prevent the loss of neutral rare alleles from the population, and increase the probability of population persistence over 50 years [27]. To date, six separate supplementations of four DFTD-affected wild devil populations have occurred, using devils sourced from a disease-free insurance metapopulation [50]. If available, genetic data should be used to select individuals for translocation to minimize inbreeding and maximize diversity [12]. Most devils in the insurance metapopulation have tissue samples taken which undergo reduced representation sequencing. These genetic data have been used to select devils for release; devils that best complement the wild sites by introducing novel alleles into the supplemented population [50]. Currently, the STDP and the University of Sydney are undertaking a large research study to investigate the impacts of these supplementations on population fitness and the prevalence of DFTD, with preliminary data showing positive signs [30]. Part of this research project includes measuring immune gene diversity, such as diversity at major histocompatibility complex (MHC) loci, in admixed devils following supplementation. Diversity at immune genes is important, because low MHC diversity in Tasmanian devils is believed to have contributed to the emergence and rapid transmission of DFTD [51].

Whilst the supplementations performed by the STDP aim to increase overall genetic variation, Kelly and Phillips (2015) suggest using a more aggressive approach, called targeted gene flow. Targeted gene flow introduces resistant alleles into a population by translocating individuals with beneficial adaptations. This technique has not been adopted into devil conservation efforts primarily because: (1) it is a relatively new concept with some aspects still speculative [52]; (2) investigations into genomic regions associated with DFTD resistance are still under investigation [53]; and (3) there are concerns that translocating individuals selected specifically for disease resistance may increase adaptability towards the disease but reduce future adaptive potential.

2.3. Benefits of Supplementing Populations

Performing supplementations to genetically manage populations threatened by disease is an emerging concept. Currently, there is little direct evidence regarding the ability of supplementations to reduce the extinction risk of diseased populations. Supplementation to preserve non-disease threatened populations [34,40–42] may provide some insight into the potential benefits of supplementing DFTD-affected devil populations.

The mountain pygmy-possum (*Barramys parvus*) is a prominent example of a threatened species, without specific known disease threats, which has benefitted from supplementation. The mountain pygmy-possum is a small, Australian marsupial, located within three, genetically distinct populations that have limited gene flow [42]. The Mount Buller population was under serious threat of extinction due to habitat clearing, which had reduced the size, genetic diversity, and fecundity of the population. To improve the demographic and genetic integrity of this population, it was supplemented with a small number of males ($n = 11$) sourced from one of the two other remaining populations, between 2011 and 2014 [44]. Following supplementation, the population exhibited a 68 percent growth in population size, increased genetic diversity, and integration of novel alleles into the gene pool over a five-year period [42]. There was evidence of hybrid fitness, seen in admixed individuals being larger and producing more offspring [42]. The response of the population to supplementation demonstrated that supplementation provided distinct genetic benefits to the Mount Buller population, as well as boosting population fitness and demographic parameters.

The response observed in the mountain pygmy-possum is indicative of genetic rescue—an increase in population fitness (growth) of a population which is suffering inbreeding depression owing to the immigration of new alleles [43]. Genetic rescue occurred, partially due to the supplementation being performed in conjunction with efforts to restore the possum's degraded habitat [42]. These actions align with IUCN guidelines that, when performing supplementations, the threat that caused the population's decline should be minimized [33].

However, a key point of comparison between supplementing disease and non-disease-affected populations is that, whilst many threats to populations can at least be partially addressed, as noted above, disease threats can be difficult to mitigate. As such, population growth in diseased populations following supplementation is likely to be slower, resulting in genetic restoration. Genetic restoration is an increase in genetic variation and relative, but not absolute, fitness owing to the immigration of new alleles [43]. Whilst genetic restoration does not necessarily produce the same demographic recovery achieved by genetic rescue, it nevertheless boosts the diversity of the population and helps ensure its future adaptive potential. Measuring genetic change in a population following supplementation (e.g., inbreeding, and neutral and functional diversity) is a useful measure of supplementation success in populations threatened by disease.

For Tasmanian devils, this means that supplementing populations under the Wild Devil Recovery Project may not lead to significant demographic recovery due to the ongoing presence of disease, but may lead to genetic benefits for populations which are showing signs of increasing inbreeding [27]. At this time, it is undetermined if these supplementations will benefit or hinder the species' ability to co-evolve with the disease. Although evidence of a selective response in devils to DFTD has been inconsistent [53,54], the suggestion of possible evolutionary change in devil populations exposed to DFTD [53] has prompted concern that sourcing the released devils from the DFTD-free insurance population will be counter-productive for the evolution of host resistance to the disease [55]. It should be noted, however, that the insurance metapopulation has been acquiring orphan devils from diseased populations since 2009 [56,57], so much post-DFTD wild diversity is likely to be represented in the insurance metapopulation.

3. Extending Principles Adopted in Tasmanian Devil Conservation Efforts to Other Populations Suffering from Disease

3.1. Genetic Management of Disease-Affected Populations

Tasmanian devils are one of the few species primarily suffering from decline due to disease that have genetic management via supplementation incorporated into their conservation program. It is unfortunate that the genetic management of disease-threatened populations has not been more widely accepted, because other species impacted by disease could benefit from supplementation. For example, mountain yellow-legged frogs (*Rana muscosa*) have declined, in part, due to chytrid fungus [58]. They are currently listed as endangered by the IUCN [59]. There are nine small populations of frogs, persisting in southern California, U.S.A., that require management to avoid extinction [60]. These populations are structured and possess low within-population variation, indicating that the populations are genetically isolated [61]. Currently, there is no evidence of inbreeding within these populations [61]; however, given that gene flow between populations is limited, it may emerge. Implementing strategies, similar to those adopted in devil conservation efforts, could avoid loss of alleles due to genetic drift and inbreeding. Captive frogs involved in San Diego Zoo's successful breeding program [61] could potentially be an appropriate source of individuals for supplementation. Augmenting gene flow may reduce extinction risk in the mountain yellow-legged frog, and other species subject to significant disease-induced population decline, by maintaining genetic diversity and providing alternate alleles upon which selection can act.

3.2. Fear of Failure

There is a hesitancy to accept supplementation as a conservation strategy even in non-diseased populations [12]. From a genetic perspective, a primary perceived risk associated with supplementation is outbreeding depression [12]. Outbreeding depression occurs when the mixing of two genetically distinct populations leads to reduced fitness in hybrid offspring, due to the disruption of co-adapted gene complexes [9,43]. However, outbreeding depression is rare [62], and mainly seen when the mixed populations are highly divergent or show high genetic structure [12,62]. An example is the Tanta Mountain ibex (*Capra ibex ibex*), where a population was supplemented with individuals sourced from two related subspecies. Hybrids of the incumbent and introduced ibex showed altered reproductive habits that led to offspring death and population extinction [63]. To combat this, a decision-making framework has been developed to predict the likelihood of outbreeding depression occurring [64]. The outcome of mixing populations cannot be definitively known until post-supplementation, and therefore there is no guarantee that a supplementation will not have negative impacts. However, existing knowledge surrounding the probability of outbreeding depression [12,62,64] and the characteristics of a suitable source population [9,35] allows for better selection of candidates for supplementation. Coupled with the ever-increasing global production of wildlife genomes and modelling methods, we are in a better position than ever to predict the impacts of supplementation on threatened populations, both with and without disease.

Outbreeding depression is not the only risk that needs to be considered when making the decision about whether to supplement disease-threatened populations. Other concerns include the loss of local adaptations, loss of species purity due to mixing of gene pools, genetic replacement, and disease spread [12,44]. In situations where the disease threat is difficult to mitigate, conservation managers may be hesitant to release healthy individuals into the wild where they may be exposed to the disease, especially if the size of the disease-free source population is small. As conducted with the Tasmanian devil, modelling and simulations, such as population viability analyses, which incorporate the demographic effects of disease, can be useful to weigh risks and options when considering supplementation [27]. In addition, wildlife disease risk analyses offer a structured method to identify, prevent, and mitigate disease risks associated with supplementation prior to implementation [65].

A well-intentioned fear of doing harm may prevent supplementation being used to genetically manage populations threatened by disease. However, a decision to not implement genetic management, or other conservation actions, is still an active conservation action with its own ramifications [12,17,36,42]. For example, the decision was made to preserve the taxonomic integrity of the dusky seaside sparrow (*Ammodramus maritimus nigrescens*) rather than outcross the remaining individuals to a closely related subspecies [66]. This decision resulted in a detrimental outcome: the species became extinct, and the unique diversity that otherwise might have been preserved was lost. This does not necessarily mean that this outcrossing event would have prevented the extinction of the dusky seaside sparrow, but we shall never know. Instead, this example encourages us to recognize that species have become extinct after active decisions not to intervene were made.

There will always be some degree of risk associated with implementing a management strategy. However, if supplementation is done to serve a conservation purpose in response to genetic isolation, inbreeding, and low genetic diversity, the rewards will often outweigh the risks, especially when facing the ultimate risk of species extinction. For a diseased population that is declining, the fear of failure needs to be weighed against the potential consequences of inbreeding, loss of genetic diversity, and decline in adaptive potential if the decision to not supplement the population is made.

4. Conclusions

There are a range of threatened species, and subsequently a range of management actions to ensure their persistence. For many species suffering infectious diseases, conservation management actions are further complicated by the disease. Here, we suggest that genetically managing disease-affected populations may assist in reducing their extinction risk when the disease threat cannot be easily mitigated. This is not a definitive solution but may buy the species time to co-evolve with their disease. As has been implemented with the Tasmanian devil, supplementation may lead to genetic restoration of these disease-threatened populations, alleviating loss of genetic diversity and maintaining their adaptive potential. Although it is possible that not all populations are suitable for this type of genetic management [64], these populations can generally be identified by preliminary screening [12]. In addition, the primary genetic risk of outbreeding depression is rare, usually manageable, and often less threatening than the extinction risk these populations face in the absence of augmented gene flow [12]. Future supplementations should be performed in conjunction with long-term monitoring to expand the existing knowledge of genetic rescue/restoration activities and gather empirical evidence of its success in diseased populations. Rather than delaying action due to a fear of provoking harmful consequences, genetic management should be recognized as a potentially beneficial conservation technique, allowing for the development of more effective management practices where action rather than inaction is favoured.

Author Contributions: Conceptualization, G.L.G., C.E.G., K.B. and C.J.H.; investigation, G.L.G.; resources, G.L.G.; writing—original draft preparation, G.L.G.; writing—review and editing, C.E.G., K.B. and C.J.H.; supervision, C.E.G., K.B. and C.J.H.; project administration, C.J.H.; funding acquisition, C.E.G., K.B. and C.J.H. All authors have read and agreed to the published version of the manuscript.

Funding: This research was funded by the Australian Research Council Linkage (LP180100244).

Acknowledgments: Thanks to members of the Australasian Wildlife Genomics Group who participated in the many discussions regarding genetic rescue in the presence of disease.

Conflicts of Interest: The authors declare no conflict of interest.

References

1. Ceballos, G.; Ehrlich, P.R.; Barnosky, A.D.; García, A.; Pringle, R.M.; Palmer, T.M. Accelerated modern human–induced species losses: Entering the sixth mass extinction. *Sci. Adv.* **2015**, *1*, e1400253. [CrossRef]
2. Díaz, S.; Fargione, J.; Chapin, F.S.; Tilman, D. Biodiversity Loss Threatens Human Well-Being. *PLoS Biol.* **2006**, *4*, e277. [CrossRef]
3. Mace, G.M.; Norris, K.; Fitter, A.H. Biodiversity and ecosystem services: A multilayered relationship. *Trends Ecol. Evol.* **2012**, *27*, 19–26. [CrossRef]
4. McKnight, D.T.; Schwarzkopf, L.; Alford, R.A.; Bower, D.S.; Zenger, K.R. Effects of emerging infectious diseases on host population genetics: A review. *Conserv. Genet.* **2017**, *18*, 1235–1245. [CrossRef]
5. De Castro, F.; Bolker, B. Mechanisms of disease-induced extinction. *Ecol. Lett.* **2005**, *8*, 117–126. [CrossRef]
6. Cunningham, A.A.; Daszak, P.; Wood, J.L.N. One Health, emerging infectious diseases and wildlife: Two decades of progress? Philosophical transactions. *Biol. Sci.* **2017**, *372*, 20160167. [CrossRef]
7. Douglass, F.J.; Harmony, J.D.; Nelson, C.D. A conceptual framework for restoration of threatened plants: The effective model of American chestnut (*Castanea dentata*) reintroduction. *New Phytol.* **2013**, *197*, 378–393.
8. Deem, S.L.; Karesh, W.B.; Weisman, W. *Putting Theory into Practice: Wildlife Health in Conservation*; Blackwell Science Inc: Boston, MA, USA, 2001; pp. 1224–1233.
9. Edmands, S. Between a rock and a hard place: Evaluating the relative risks of inbreeding and outbreeding for conservation and management. *Mol. Ecol.* **2006**, *16*, 463–475. [CrossRef]
10. Huisman, J.; Kruuk, L.E.B.; Ellis, P.A.; Clutton-Brock, T.H.; Pemberton, J.M. Inbreeding depression across the lifespan in a wild mammal population. *Proc. Natl. Acad. Sci. USA* **2016**, *113*, 3585–3590. [CrossRef] [PubMed]

11. Wisely, S.M.; Ryder, O.A.; Santymire, R.M.; Engelhardt, J.F.; Novak, B.J. A Road Map for 21st Century Genetic Restoration: Gene Pool Enrichment of the Black-Footed Ferret. *J. Hered.* **2015**, *106*, 581–592. [CrossRef] [PubMed]

12. Ralls, K.; Ballou, J.D.; Dudash, M.R.; Eldridge, M.D.B.; Fenster, C.B.; Lacy, R.C.; Sunnucks, P.; Frankham, R. Call for a Paradigm Shift in the Genetic Management of Fragmented Populations. *Conserv. Lett.* **2017**, *11*, e12412. [CrossRef]

13. Hamish, M.; Andy, D. Disease, habitat fragmentation and conservation. *Proc. R. Soc. B Biol. Sci.* **2002**, *269*, 2041–2049.

14. Joseph, M.B.; Mihaljevic, J.R.; Arellano, A.L.; Kueneman, J.G.; Preston, D.L.; Cross, P.C.; Johnson, P.T.J. Taming wildlife disease: Bridging the gap between science and management. *J. Appl. Ecol.* **2013**, *50*, 702–712. [CrossRef]

15. MacInnes, C.D.; Smith, S.M.; Tinline, R.; Ayers, N.R.; Bachmann, P.; Ball, D.G.A.; Calder, L.A.; Crosgrey, S.J.; Fielding, C.; Hauschildt, P.; et al. Elimination of rabies from red foxes in eastern ontario. *J. Wildl. Dis.* **2001**, *37*, 119–132. [CrossRef] [PubMed]

16. Portier, J.; Ryser-Degiorgis, M.-P.; Hutchings, M.R.; Monchâtre-Leroy, E.; Richomme, C.; Larrat, S.; Van Der Poel, W.H.M.; Dominguez, M.; Linden, A.; Santos, P.T.; et al. Multi-host disease management: The why and the how to include wildlife. *BMC Vet. Res.* **2019**, *15*, 1–11. [CrossRef] [PubMed]

17. Scheele, B.C.; Hunter, D.A.; Grogan, L.F.; Berger, L.; Kolby, J.E.; McFadden, M.S.; Marantelli, G.; Skerratt, L.F.; Driscoll, D.A. Interventions for Reducing Extinction Risk in Chytridiomycosis-Threatened Amphibians. *Conserv. Biol.* **2014**, *28*, 1195–1205. [CrossRef]

18. Scheele, B.C.; Foster, C.N.; Hunter, D.A.; Lindenmayer, D.B.; Schmidt, B.R.; Heard, G.W. Living with the enemy: Facilitating amphibian coexistence with disease. *Biol. Conserv.* **2019**, *236*, 52–59. [CrossRef]

19. Boots, M.; Best, A.; Miller, M.R.; White, A. The role of ecological feedbacks in the evolution of host defence: What does theory tell us? *Philos. Trans. R. Soc. B Biol. Sci.* **2008**, *364*, 27–36. [CrossRef]

20. Jachowski, D.S.; Gitzen, R.A.; Grenier, M.B.; Holmes, B.; Millspaugh, J.J. The importance of thinking big: Large-scale prey conservation drives black-footed ferret reintroduction success. *Biol. Conserv.* **2011**, *144*, 1560–1566. [CrossRef]

21. Pearse, A.M.; Swift, K. Allograft theory: Transmission of devil facial-tumour disease. *Nature* **2006**, *439*, 549. [CrossRef]

22. Pye, R.J.; Pemberton, D.; Tovar, C.; Tubio, J.M.C.; Dun, K.A.; Fox, S.; Darby, J.; Hayes, D.; Knowles, G.W.; Kreiss, A.; et al. A second transmissible cancer in Tasmanian devils. *Proc. Natl. Acad. Sci. USA* **2015**, *113*, 374–379. [CrossRef]

23. Hawkins, C.; Baars, C.; Hesterman, H.; Hocking, G.; Jones, M.; Lazenby, B.; Mann, D.; Mooney, N.; Pemberton, D.; Pyecroft, S.; et al. Emerging disease and population decline of an island endemic, the *Tasmanian devil Sarcophilus harrisii. Biol. Conserv.* **2006**, *131*, 307–324. [CrossRef]

24. Pye, R.; Hamede, R.; Siddle, H.V.; Caldwell, A.; Knowles, G.W.; Swift, K.; Kreiss, A.; Jones, M.E.; Lyons, A.B.; Woods, G.M. Demonstration of immune responses against devil facial tumour disease in wild *Tasmanian devils. Biol. Lett.* **2016**, *12*, 20160553. [CrossRef] [PubMed]

25. Lazenby, B., Tobler, M.W.; Brown, W.E.; Hawkins, C.E.; Hocking, G.J.; Hume, F.; Huxtable, S.; Iles, P.; Jones, M.E.; Lawrence, C.; et al. Density trends and demographic signals uncover the long-term impact of transmissible cancer in *Tasmanian devils. J. Appl. Ecol.* **2018**, *55*, 1368–1379. [CrossRef]

26. Jones, M.E.; Cockburn, A.; Hamede, R.; Hawkins, C.; Hesterman, H.; Lachish, S.; Mann, D.; McCallum, H.; Pemberton, D. Life-history change in disease-ravaged *Tasmanian devil* populations. *Proc. Natl. Acad. Sci. USA* **2008**, *105*, 10023–10027. [CrossRef]

27. Grueber, C.E.; Fox, S.; McLennan, E.A.; Gooley, R.M.; Pemberton, D.; Hogg, C.J.; Belov, K. Complex problems need detailed solutions: Harnessing multiple data types to inform genetic management in the wild. *Evol. Appl.* **2018**, *12*, 280–291. [CrossRef] [PubMed]

28. Hawkins, C.E.; McCallum, H.; Mooney, N.; Jones, M.; Holdsworth, M. *Sarcophilus Harrisii;* The IUCN Red List of Threatened Species: Cambridge, UK, 2008.

29. Pye, R.; Patchett, A.; McLennan, E.; Thomson, R.; Carver, S.; Fox, S.; Pemberton, D.; Kreiss, A.; Morelli, A.B.; Silva, A. Immunization strategies producing a humoral IgG immune response against devil facial tumor disease in the majority of *Tasmanian devils* destined for wild release. *Front. Immunol.* **2018**, *9*, 259. [CrossRef]

30. Fox, S.; Seddon, P.J. Wild devil recovery: Managing devils in the presence of disease. In *Saving the Tasmanian Devil Recovery through Science-Based Management*; Hogg, C., Fox, S., Pemberton, D., Belov, K., Eds.; CSIRO: Victoria, Australia, 2019.

31. Lachish, S.; Miller, K.J.; Storfer, A.; Goldizen, A.W.; Jones, M.E. Evidence that disease-induced population decline changes genetic structure and alters dispersal patterns in the *Tasmanian devil. Heredity* **2010**, *106*, 172–182. [CrossRef]

32. Miller, W.; Hayes, V.M.; Ratan, A.; Petersen, D.C.; Wittekindt, N.E.; Miller, J.; Walenz, B.; Knight, J.; Qingyu, W.; Zhao, F.; et al. Genetic diversity and population structure of the endangered marsupial *Sarcophilus harrisii (Tasmanian devil). Proc. Natl. Acad. Sci. USA* **2011**, *108*, 12348–12353. [CrossRef]

33. IUCN. *Guidlines for Reintroductions and Other Conservation Translocations*; Commission, I.S.S., Ed.; IUCN: Gland, Switzerland, 2013.

34. Hogg, J.T.; Forbes, S.H.; Steele, B.M.; Luikart, G. Genetic rescue of an insular population of large mammals. *Proc. R. Soc. B Boil. Sci.* **2006**, *273*, 1491–1499. [CrossRef]

35. Kronenberger, J.A.; Gerberich, J.C.; Fitzpatrick, S.W.; Broder, E.D.; Angeloni, L.M.; Funk, W.C. An experimental test of alternative population augmentation scenarios. *Conserv. Biol.* **2018**, *32*, 838–848. [CrossRef]

36. Stowell, S.M.L.; Pinzone, C.A.; Martin, A.P. Overcoming barriers to active interventions for genetic diversity. *Biodivers. Conserv.* **2017**, *16*, 613–1765. [CrossRef]

37. Robinson, Z.L.; Coombs, J.A.; Hudy, M.; Nislow, K.H.; Letcher, B.H.; Whiteley, A.R. Experimental test of genetic rescue in isolated populations of brook trout. *Mol. Ecol.* **2017**, *26*, 4418–4433. [CrossRef]

38. Hedrick, P.W.; Garcia-Dorado, A. Understanding Inbreeding Depression, Purging, and Genetic Rescue. *Trends Ecol. Evol.* **2016**, *31*, 940–952. [CrossRef]

39. Heber, S.; Varsani, A.; Kuhn, S.; Girg, A.; Kempenaers, B.; Briskie, J. The genetic rescue of two bottlenecked South Island robin populations using translocations of inbred donors. *Proc. Biol. Sci.* **2013**, *280*, 20122228. [CrossRef]

40. Hedrick, P.W.; Fredrickson, R. Genetic rescue guidelines with examples from Mexican wolves and Florida panthers. *Conserv. Genet.* **2009**, *11*, 615–626. [CrossRef]

41. Johnson, W.E.; Onorato, D.; Roelke, M.E.; Land, E.D.; Cunningham, M.; Belden, R.C.; McBride, R.; Jansen, D.; Lotz, M.; Shindle, D.; et al. Genetic Restoration of the Florida Panther. *Science* **2010**, *329*, 1641–1645. [CrossRef] [PubMed]

42. Weeks, A.R.; Heinze, D.; Perrin, L.; Stoklosa, J.; Hoffmann, A.A.; Van Rooyen, A.; Kelly, T.; Mansergh, I. Genetic rescue increases fitness and aids rapid recovery of an endangered marsupial population. *Nat. Commun.* **2017**, *8*, 1–6. [CrossRef] [PubMed]

43. Whiteley, A.R.; Fitzpatrick, S.W.; Funk, W.C.; Tallmon, D.A. Genetic rescue to the rescue. *Trends Ecol. Evol.* **2015**, *30*, 42–49. [CrossRef] [PubMed]

44. Weeks, A.R.; Sgro, C.M.; Young, A.G.; Frankham, R.; Mitchell, N.J.; Miller, K.A.; Byrne, M.; Coates, D.J.; Eldridge, M.D.B.; Sunnucks, P.; et al. Assessing the benefits and risks of translocations in changing environments: A genetic perspective. *Evol. Appl.* **2011**, *4*, 709–725. [CrossRef] [PubMed]

45. Hufbauer, R.; Szűcs, M.; Kasyon, E.; Youngberg, C.; Koontz, M.J.; Richards, C.M.; Tuff, T.; Melbourne, B.A. Three types of rescue can avert extinction in a changing environment. *Proc. Natl. Acad. Sci. USA* **2015**, *112*, 10557–10562. [CrossRef] [PubMed]

46. Robinson, J.; Brown, C.; Kim, B.Y.; Lohmueller, K.E.; Wayne, R.K. Purging of Strongly Deleterious Mutations Explains Long-Term Persistence and Absence of Inbreeding Depression in Island Foxes. *Curr. Biol.* **2018**, *28*, 3487–3494.e4. [CrossRef]

47. Robinson, J.A.; Vecchyo, D.O.-D.; Fan, Z.; Kim, B.Y.; Vonholdt, B.M.; Marsden, C.D.; Lohmueller, K.E.; Wayne, R.K. Genomic Flatlining in the Endangered Island Fox. *Curr. Biol.* **2016**, *26*, 1183–1189. [CrossRef] [PubMed]

48. Ralls, K.; Sunnucks, P.; Lacy, R.C.; Frankham, R. Genetic rescue: A critique of the evidence supports maximizing genetic diversity rather than minimizing the introduction of putatively harmful genetic variation. *Biol. Conserv.* **2020**, *251*, 108784. [CrossRef]

49. Gandon, S.; Michalakis, Y. Local adaptation, evolutionary potential and host-parasite coevolution: Interactions between migration, mutation, population size and generation time. *J. Evol. Biol.* **2002**, *15*, 451–462. [CrossRef]

50. Hogg, C.; McLennan, E.; Wise, P.; Lee, A.; Pemberton, D.; Fox, S.; Belov, K.; Grueber, C. Preserving the demographic and genetic integrity of a single source population during multiple translocations. *Biol. Conserv.* **2020**, *241*, 108318. [CrossRef]

51. Peel, E.; Belov, K. Lessons learnt from the Tasmanian devil facial tumour regarding immune function in cancer. *Mamm. Genome* **2018**, *29*, 731–738. [CrossRef]

52. Kelly, E.; Phillips, B.L. Targeted gene flow for conservation. *Conserv. Biol.* **2016**, *30*, 259–267. [CrossRef]

53. Epstein, B.; Jones, M.; Hamede, R.; Hendricks, S.; McCallum, H.; Murchison, E.P.; Schönfeld, B.; Wiench, C.; Hohenlohe, P.; Storfer, A. Rapid evolutionary response to a transmissible cancer in Tasmanian devils. *Nat. Commun.* **2016**, *7*, 1–7. [CrossRef]

54. Brüniche-Olsen, A.; Austin, J.J.; Jones, M.E.; Holland, B.R.; Burridge, C.P. Detecting Selection on Temporal and Spatial Scales: A Genomic Time-Series Assessment of Selective Responses to Devil Facial Tumor Disease. *PLoS ONE* **2016**, *11*, e0147875. [CrossRef]

55. Hohenlohe, P.A.; McCallum, H.I.; Jones, M.E.; Lawrance, M.F.; Hamede, R.K.; Storfer, A. Conserving adaptive potential: Lessons from Tasmanian devils and their transmissible cancer. *Conserv. Genet.* **2019**, *20*, 81–87. [CrossRef]

56. Hogg, C.J.; Lee, A.V.; Hibbard, C.J. Managing a metapopulation: Intensive to wild and all the places in between. In *Saving the Tasmanian devil: Recovery through Science-Based Management*; CSIRO: Victoria, Australia, 2019.

57. Hogg, C.J.; Lee, A.V.; Srb, C.; Hibbard, C. Metapopulation management of an Endangered species with limited genetic diversity in the presence of disease: The *Tasmanian devil Sarcophilus harrisii*. *Int. Zoo Yearb.* **2016**, *51*, 137–153. [CrossRef]

58. Gary, M.F.; Green, D.E.; Joyce, E.L. Oral Chytridiomycosis in the Mountain Yellow-Legged Frog (*Rana muscosa*). *Copeia* **2001**, *2001*, 945–953.

59. Hammerson, G. *Rana Muscosa*; The IUCN Red List of Threatened Species: Cambridge, UK, 2008.

60. Backlin, A.R.; Hitchcock, C.J.; Gallegos, E.A.; Yee, J.L.; Fisher, R.N. The precarious persistence of the Endangered Sierra Madre yellow-legged frog Rana muscosa in southern California, USA. *Oryx* **2013**, *49*, 157–164. [CrossRef]

61. Schoville, S.D.; Tustall, T.S.; Vredenburg, V.T.; Backlin, A.R.; Gallegos, E.; Wood, D.A.; Fisher, R.N. Conservation genetics of evolutionary lineages of the endangered mountain yellow-legged frog, *Rana muscosa* (*Amphibia: Ranidae*), in southern California. *Biol. Conserv.* **2011**, *144*, 2031–2040. [CrossRef]

62. Byrne, P.G.; Silla, A.J. An experimental test of the genetic consequences of population augmentation in an amphibian. *Conserv. Sci. Pract.* **2020**, *2*, 1–11. [CrossRef]

63. Templeton, A.; Hemmer, H.; Mace, G.M.; Seal, U.S.; Shields, W.M.; Woodruff, D.S. Local adaptation, coadaptation, and population boundaries. *Zoo Biol.* **1986**, *5*, 115–125. [CrossRef]

64. Frankham, R.; Ballou, J.D.; Eldridge, M.D.B.; Lacy, R.C.; Ralls, K.; Dudash, M.R.; Fenster, C.B. Predicting the Probability of Outbreeding Depression. *Conserv. Biol.* **2011**, *25*, 465–475. [CrossRef] [PubMed]

65. Hartley, M.; Sainsbury, A. Methods of Disease Risk Analysis in Wildlife Translocations for Conservation Purposes. *EcoHealth* **2017**, *14*, 16–29. [CrossRef]

66. Zink, R.M.; Kale, H.W. Conservation genetics of the extinct dusky seaside sparrow Ammodramus maritimus nigrescens. *Biol. Conserv.* **1995**, *74*, 69–71. [CrossRef]

Communication

Pug-Headedness Anomaly in a Wild and Isolated Population of Native Mediterranean Trout *Salmo trutta* L., 1758 Complex (Osteichthyes: Salmonidae)

Francesco Palmas [1,*], **Tommaso Righi** [2], **Alessio Musu** [1], **Cheoma Frongia** [1], **Cinzia Podda** [1], **Melissa Serra** [1], **Andrea Splendiani** [2], **Vincenzo Caputo Barucchi** [2] and **Andrea Sabatini** [1]

[1] Dipartimento di Scienze della Vita e dell'Ambiente, Università di Cagliari, Via T. Fiorelli 1, 09126 Cagliari (CA), Italy; alessio.musu91@gmail.com (A.M.); cheomaf@tiscali.it (C.F.); cpodda@unica.it (C.P.); melissaserra191@gmail.com (M.S.); asabati@unica.it (A.S.)

[2] Dipartimento di Scienze della Vita e dell'Ambiente, Università Politecnica delle Marche, via Brecce Bianche, 60100 Ancona, Italy; t.righi@pm.univpm.it (T.R.); andreasplendiani@hotmail.com (A.S.); v.caputo@univpm.it (V.C.B.)

* Correspondence: fpalmas@unica.it; Tel.:+39-070-678-8010

Received: 28 August 2020; Accepted: 12 September 2020; Published: 15 September 2020

Abstract: Skeletal anomalies are commonplace among farmed fish. The pug-headedness anomaly is an osteological condition that results in the deformation of the maxilla, pre-maxilla, and infraorbital bones. Here, we report the first record of pug-headedness in an isolated population of the critically endangered native Mediterranean trout *Salmo trutta* L., 1758 complex from Sardinia, Italy. Fin clips were collected for the molecular analyses (D-loop, *LDH-C1** locus. and 11 microsatellites). A jaw index (JI) was used to classify jaw deformities. Ratios between the values of morphometric measurements of the head and body length were calculated and plotted against values of body length to identify the ratios that best discriminated between malformed and normal trout. Haplotypes belonging to the AD lineage and the genotype *LDH-C1*100/100* were observed in all samples, suggesting high genetic integrity of the population. The analysis of 11 microsatellites revealed that observed heterozygosity was similar to the expected one, suggesting the absence of inbreeding or outbreeding depression. The frequency of occurrence of pug-headedness was 12.5% (two out of 16). One specimen had a strongly blunted forehead and an abnormally short upper jaw, while another had a slightly anomaly asymmetrical jaw. Although sample size was limited, variation in environmental factors during larval development seemed to be the most likely factors to trigger the deformities.

Keywords: small isolated population; Mediterranean native trout; morphological deformities

1. Introduction

Skeletal anomalies are commonplace characteristic in farmed fish all over the world [1–3]. In contrast, naturally originated malformations show a smaller incidence in wild fish populations due to the decreased viability of the abnormal fish in natural habitats [4,5]. The extent of the skeletal deformities in fish species has been reported to affect different anatomical body parts, such as the vertebral column, fins, and skull [1,6,7]. In particular, skull malformations involve mainly the splanchnocranium, hyoid arch, and gill cover [1,3].

Among these deformities, the pug-headedness anomaly is an osteological condition that results in the deformation of the maxilla, pre-maxilla, infraorbital bones, and ethmoid region. This condition determines bulging eyeballs, acutely steep foreheads, and incomplete closure of the mouth due to projection of the lower jaw [8,9]. Pug-headedness can lead to starvation and rising mortality, due to the jaw deformity, especially during the larval stages [10]. Several causes have been suggested to explain the pug-headedness anomaly, including low genetic variability and epigenetic factors, such as embryonic development disorders and aberrations induced by environmental factors variation [11,12]. Among these, traumatic shock caused by daily variations in temperature, light cycles, salinity, and dissolved oxygen concentration during the early development seem to be the most important factors [2,9]. A higher incidence of pug-headedness has been found in polluted waters [12,13]. The pug-headedness deformity is widely documented in different captive fish species of both marine and freshwater habitats [14–18], but it is rare among adults in wild populations [19,20]. Among natural populations, jaw deformities are mostly restricted to different families of marine fish species, such as Moronidae [21], Pomacanthidae [22], Rachycentridae [23], Sparidae [24], Epinephelidae, and Cichlidae [20,25].

The pug-headedness malformation has also been reported in wild salmonid populations in only a few known cases [26]. To the best of our knowledge, this is the first documented occurrence of pug-headedness in a native Mediterranean trout population. Although the taxonomy of the Mediterranean trout has not yet been resolved [27], the species is listed with the name of *Salmo cettii* Rafinesque, 1810, and is considered to be critically endangered by the International Union for Conservation of Nature [28].

In particular, in this paper we report (1) a brief morphological characterisation of the head deformities, (2) the genetic characterisation of the population using nuclear (*LDH-C1** and 11 microsatellites) and mitochondrial (entire mtDNA control region, ~1 Kb) markers, and (3) a comparison of malformed and normal specimens.

2. Materials and Methods

During the monitoring programme for the compilation of the official Fish Inventory of the Sardinia Region [29], among 116 native trout analysed in 10 rivers, two malformed specimens were found only in the Furittu Stream in June 2017 and June 2019 (39°26.248′ N, 9°22.036′ E) (Figure 1). The stream is located in the south-eastern sector of Sardinia, running for about 10 km to merge with the River S'Acqua Callenti to form the Flumendosa River. The Furittu stream is located in the mountainous area of Monte Genis, which is an area with very low anthropogenic pressure, is hard to reach and is not subjected to wind coming from polluted areas. The Furittu stream watershed area is 21 km^2, and the land use is largely natural and composed of forests, Mediterranean shrubland and very little pasture land. Bi-seasonal climatic features, with hot arid summers and a rainy autumn/winter season, determine a periodic of hydrographic isolation (up to four months per year) for the upper part of the stream. The stream is 0.5–4.5 m wide and 0.10–2.00 m deep with an average slope of 3.8% throughout its whole length. It has the typical geomorphologic characteristics of a Mediterranean mountain stream, with a complex and fragmented mesohabitat dominated by pools (72%), riffles (20%) and small cascades (8%). The streambed consists of rocks, boulders, gravel, rubble, and woody debris, with riparian vegetation composed of trees adjacent to the stream characterised by holm oak (*Quercus ilex*) and oleander (*Nerium oleander*).

A total of 16 trout were captured in the upper region of the stream using low-frequency, pulsed DC electrofishing and stored in cool, aerated water. All specimens were measured for total weight (TW, g) and total length (TL, cm), placed in a narrow transparent tank filled with water, and photographed from the left side. From each fish, a small fin clip was removed and conserved in 95% ethanol until DNA extraction. After processing, the fish were placed in large containers and released in the stream. Estimates of the total number of fish in a 100 m section of stream and trout estimated densities (N fish/m^2) were obtained using two-pass depletion method [30]. The Sardinian specimens were compared with another trout from Neia River (42°38.31′ N, 13°14.96′ E; Tronto basin of Italian Apennine area) that showed similar jaw anomalies.

Figure 1. Geographic position of the sampling station in Furittu Stream, south-eastern Sardinia (Italy). Lower right panel: location of Furittu stream (■) and Neia River (●).

Genetic analyses were performed by using both mitochondrial (D-loop) and nuclear markers (*LDH-C1** and 11 microsatellite loci). The entire mitochondrial control region (D-loop) was PCR amplified and sequenced using the primers 28RIBa [31] and HN20 [32]. PCR amplifications were performed in 20 µL of reaction mixtures (approximately 80 ng of template DNA, 0.15 units MyTaq polymerase (Bioline GmbH, Luckenwalde, Germany), 1X MyTaq Buffer (Bioline GmbH, Luckenwalde, Germany) and 5 pmol of each primer) for 30 cycles (95 °C, 45 s; 53 °C, 30 s; 72 °C 90 s). Cycling was preceded by a 3 min denaturing step at 95 °C and followed by a 7 min final extension at 72 °C. The mtDNA sequences were aligned using the computer program ClustalW [33] and compared with reference *S. trutta* CR sequences from GenBank using BLASTn (http://blast.ncbi.nlm.nih.gov/Blast.cgi). The specimens were genotyped at the *LDH-C1** gene, coding the LDH enzyme [34].

A 440 bp segment of the *LDH-C1** nuclear locus was PCR-amplified with the primers Ldhxon3F and Ldhxon4R [34] and digested with a BseLI restriction enzyme. *LDH-C1** allows the discrimination of north-western European populations, characterised by the *90 allele, from native Mediterranean population (*100 allele), and hybrids that present both alleles [34]. PCR amplifications were performed in 25 µL of reaction mixtures (approximately 200 ng of template DNA, 0.15 units MyTaq polymerase (Bioline GmbH, Luckenwalde, Germany), 1× MyTaq Buffer (Bioline GmbH, Luckenwalde, Germany)and 5 pmol of each primer) for 30 cycles (95 °C, 60 s; 64 °C, 60 s; 72 °C 60 s). Cycling was preceded by a 5-min denaturation step at 95 °C and followed by a 10-min final extension at 72 °C. Amplicon digestion was performed using BseLI restriction enzyme (Thermo Fisher Scientific, Waltham, MA, USA) following the manufacturers' protocol and visualised on 2% agarose gel. Following [35], eleven non-coding microsatellite loci were labelled with fluorescent dye and multiplexed in two separate reactions. PCR amplifications were performed in 15 µL of reaction mixtures containing approximately 80 ng of template DNA, 0.4 unit MyTaq polymerase (Bioline GmbH, Luckenwalde, Germany), 1X MyTaq Buffer (Bioline GmbH, Luckenwalde, Germany), and 2.5 pmol of each primer. A Touchdown protocol was used to optimise the amplification. PCR amplicons were electrophoresed using an ABI-PRISM 3130xl Genetic Analyzer (Applied Biosystems, Foster City, CA, USA) [36]. ARLEQUIN 3.5.1.3 [37] was used to calculate the observed (Ho) and expected (He) heterozygosity and deviations from the Hardy–Weinberg equilibrium, while the inbreeding coefficient (F_{IS}) was estimated in FSTAT [38]. The software COLONY software [39] was used to determine family structure in the Furittu stream population (length of run = 3, level of precision likelihood = high). The pair likelihood score (PLS) /full likelihood (FL) combined (= FPLS) algorithm was selected to establish only full-sibs listings. Finally, individual admixture coefficient (*q*) were estimated for Furittu population using the software STRUCTURE [40] A domestic sample from two hatcheries was used for comparison. The domestic ancestry in Furittu population was calculated performing a run assuming a K = 2 (i.e., domestic vs. native) adopting a burn-in of 1,000,000 iterations, followed by 500,000 iterations leaving the other parameters as default.

To verify jaw deformities, we used a modified version of the jaw index (JI) [41] (Figure 2a). We categorised the specimens to different degrees of upper jaw index (UJI) as follow: values above 1 were categorised as normal jaw deformity (NJ), from 0.99 to 0.90 as mildly jaw-deformed (MJD) and below 0.89 severely jaw-deformed (SJD).

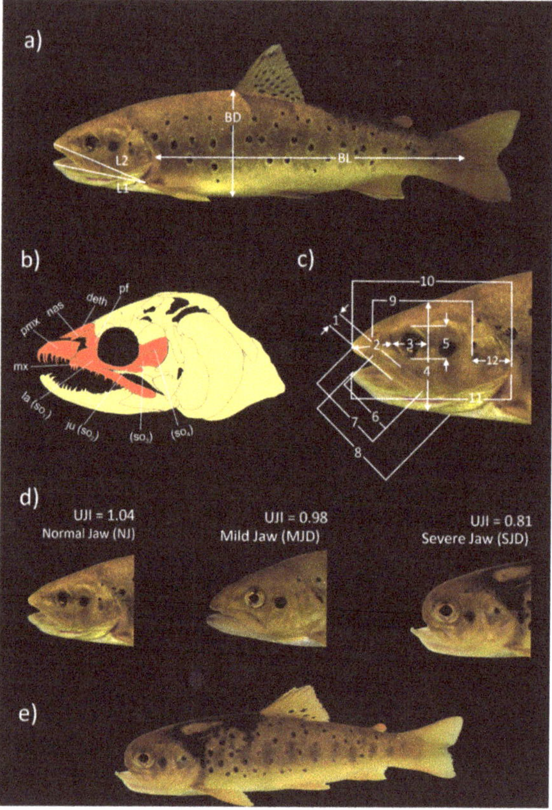

Figure 2. Specimens from Furittu stream. (**a**) Locations of L1 and L2 measurements used to calculate the upper jaw index (UJI), body length (BL) and body depth (BD), (**b**) skull of trout and in red bones involved in the pughead deformity (pmx = premaxilla; mx = maxilla; deth = dermethmoid; nas = nasal; pf = prefrontal; la = lacrimal (suborbital 1); ju = jugal (suborbital 2); so3 = suborbital 3; so4 = suborbital 4), (**c**) measurements taken for morphometric comparison (Table S1, Supplementary Materials), (**d**) examples of normal jaw (NJ), middle jaw deformity (MJD) and severe jaw deformity (SJD). Pughead specimen of Mediterranean trout (**e**).

Since the affected species often show other skull bones deformities (Figure 2b), morphological variables of the head [42] (Figure 2c) were collected from each trout after setting the scale factor using TPSDIG2 v2.31 [43]. Ratios between values of morphometric measurements of the head and body length (BL) were calculated and plotted against values of BL, to identify the ratios that best discriminated between the malformed (MJD, SJD) and normal trout (NJ). To estimate the changes in nutritional conditions among normal and deformed specimens, we used the residuals analysis in linear regression between LT and TW after logarithmic transformation of the data [17].

3. Results

In the Furittu stream, a total of 16 adult trout were examined (12.5–25.5 cm and 24.9–183.1 g) (Table S1, Supplementary Materials). The estimates of the total number of fish in the 100 m of section varied between 15 ± 4 and 18 ± 0.5 in June 2017 and 2018, respectively. Estimated densities were relatively low in both campaigns and ranged from 0.06 ind/m^2 to 0.09 ind/m^2.

The incidence rate of jaw deformities of Furittu stream trout was 12.5% (2 of 16 native trout, 95% confidence interval [CI] = 1.6–38.6%). One specimen, which showed a strongly blunted forehead and an abnormally short upper jaw (UJI = 0.81) was categorised as SJD, while the another one manifested a slightly asymmetrical jaw (UJI = 0.98, categorised as MJD). The remaining specimens showed values of UJI above 1 and were considered normal (NJ) (Figure 2d). The Neia trout with an UJI of 0.95 was categorised as MJD (Table S1, Supplementary Materials).

The mtDNA analysis showed that all Sardinian trout shared the same haplotype (AD-Tyrr7) belonging to the AD lineage (Adriatic) (sensu [44]), which was observed for the first time in this population. The new sequence was deposited in GenBank under accession number MT503201. In all samples, we observed the genotype *LDH-C1*100/100* fixed in the Mediterranean native population. The trout population from the Neia River (central Apennine) showed high frequency of the non-native allele *LDH-C1* 90* (0.77) and the presence of AT non-native mitochondrial lineage with a frequency of 0.23. The 11 microsatellites showed higher levels of observed heterozygosity in the Neia population (0.80) compared to the Furittu stream population (0.50) (Table 1). In Furittu stream, the mean *q* values were close to 1 with a very narrow CI (mean = 0.99, CI= 0.97–1).

Table 1. Mitochondrial lineage and *LDH-C1** observed frequencies, observed (H$_O$) and expected (H$_E$) heterozygosity and inbreeding coefficient (F$_{IS}$) estimated from 11 microsatellite loci in the Furittu and Neia populations. AD (Adriatic), ME (Mediterranean) and AT (Atlantic) lineages of *Salmo trutta*. Standard deviation (s.d.).

River	Sample Size	MtDNA			LDH-C1*		Microsatellite		F$_{IS}$
		AD	ME	AT	*90	*100	H$_O$ (s.d.)	H$_E$ (s.d.)	
Furittu	16	1.00				1.00	0.50 (0.26)	0.46 (0.14)	−0.089
Neia	22	0.45	0.32	0.23	0.77	0.23	0.80 (0.18)	0.83 (0.20)	0.051

No significant departure from the Hardy–Weinberg equilibrium was found in the studied populations. Analysis of family structure suggested that the Furittu stream population was composed of eight families of generally one or two individuals (Table S2, Supplementary Materials). The two abnormal specimens were found to be unrelated.

In the Furittu stream, one trout showed anomalies in skull measurements typical of pug-headedness osteological malformation (SJD), with a shortened neurocranium and upper jaw (Figure 2e), while another specimen (MJD) had slightly shorter upper jaw and greater head. In particular, the pug-headed trout (SJD) showed a greater depth of the head (DH) (Figure 3a), while the specimen with the middle jaw deformity (MJD) had the longest head (LHL) (Figure 3b). Surprisingly, the SJD individual also showed a longer LHL compared to normal specimens (NJ) (Figure 3b). The longer LHL found in the specimen affected by the pug-headedness (SJD) was thought to be a consequence of a slightly greater operculum length (LO) (Figure 3c). In the SJD specimens, the snout was almost absent due to the curving of the ethmoid region (deth, nas), and maxillary bones (mx) (Figure 2b). These malformations were confirmed by the smallest snout (LS) in SJD, while the MJD trout showed the longest LS (Figure 3d). The anomalies of infraorbital bones (pf, la, ju, so3, so4) in SJD trout determined larger and bulbed eyeballs compared to MJD and NJ specimens (Figure 3e,f) despite the fact that the ratio between HO and VO didn't show differences from the regular circular shape of infraorbital bones. As shown in the MJD specimen, the deformed trout from the Neia River exhibited a slightly prominent lower jaw in comparison to normal trout from Sardinia (Figure 3b).

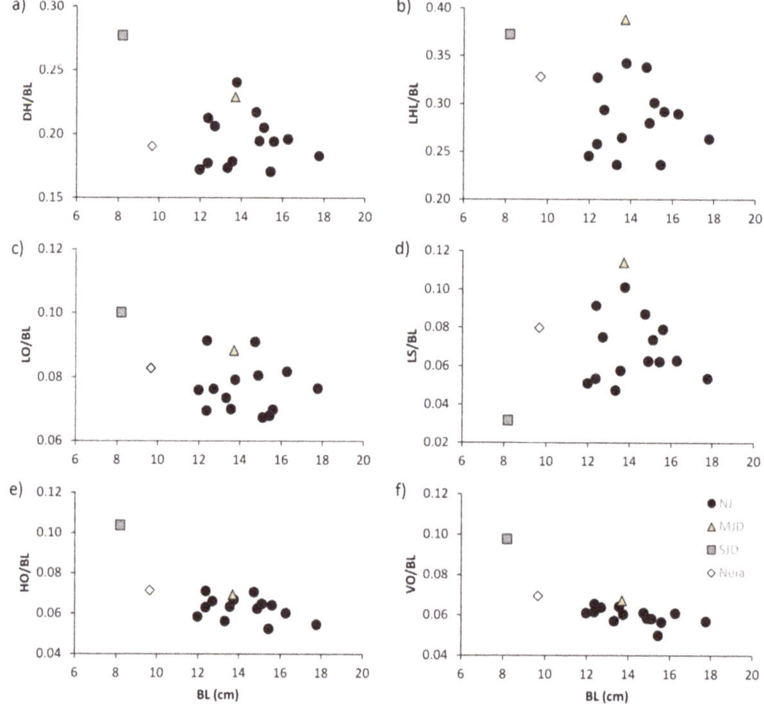

Figure 3. Relationship between body length (BL) and the ratios among values of head depth (DH) (**a**), lower head length (LHL) (**b**), operculum length (LO) (**c**), snout length (SL) (**d**), horizontal orbital length (HO) (**e**) and vertical orbital length (VO) (BL) for specimens with normal jaw deformity (NJ), middle jaw deformity (MJD) and severe jaw deformity (SJD) from Furittu stream and one specimen from Neia river (**f**).

Despite this deformity, the residuals generated by linear regression (LT/TW) revealed no relevant differences between normal and deformed specimens, indicating that the deformed trout were robust and healthy (Table S1, Supplementary Materials).

4. Discussion

The pug-headedness deformity in adults wild brown trout was first described in 1929 by the American ichthyologist Eugene Wills Gudgers [26]. Here, we present the first scientific report of head skeletal deformities in a wild population of Mediterranean native trout and one of the few cases reported for adult specimens of the genera *Salmo*.

In the Furittu stream, the head malformations were observed in two specimens out of 16 Mediterranean trout (12.5% of occurrence) captured during two sampling campaigns conducted in June 2017 and June 2018. One specimen (SJD) showed the typical malformation of pug-headness, with the antero-posterior compression of the ethmoid region and upper jaw, while the other specimen (MJD) had a slightly shorter upper jaw and a longer head. The Neia trout (MJD) also exhibited an asymmetry between the lower and upper jaw.

In unpolluted natural habitats, the occurence of pug-headedness in other genera is generally less than 1% [45]. The occurrence of head deformities in the Furittu Stream exceeded the rates observed in the polluted habitats (from 0.5% to 3%) [13], while it was comparable with that of hatchery-reared fish larvae, which have a much greater frequency of skeletal skull abnormalities, including pug-headedness, compared to their wild populations [1,3,46].

The periodic hydrographic isolation that occurs in the downstream part of the river could represent an impassable barrier that prevents the connectivity with domestic trout that have been stocked in the Flumendosa River since the beginning of the 1960s. In fact, the trout population of the Furittu stream is immune by genetic introgression with non-native traits, as detected by the genetic analysis. Before this analysis, in Sardinia, ancestral trout populations (AD lineage, *sensu* [44]) were found in two watersheds of southern Sardinia (Cixerri and Pula basins) [47–50]. Although these populations exist in small headwater habitats isolated by artificial or natural barriers, morphological analysis revealed no presence of trout with head deformations [49]. Furthermore, in a comparative study in morphological characterisation of Corsican and Sardinian trout, it was highlighted that relatively larger heads are present in native trout populations compared to Atlantic and *S. macrostigma* specimens, while no head skeletal aberrations were found [51].

Even though many studies have reported a wide range of factors that trigger such deformities, the exact cause may be difficult to determine. Possible sources of such an aberration could include a wide range of epigenetic factors, such as temperature and oxygen fluctuation during egg incubation [9,52], prenatal stress of mature females induced by environmental changes [53], influences of diet composition on larval phases [54,55], and environmental pollution [13,14]. However, endogamy in small populations is indicated as the most likely factor to trigger the deformity [17]. Low genetic variability, as a result of inbreeding depression, is known to occur in fragmented and small salmonid populations [56,57]. Salmonid populations with morphological deformities, due to the loss of genetic variation and inbreeding depression, show a lower survival rate compared to the normal conspecific populations [58,59]. In this context, similar morphological deformities were detected in two watersheds showing significant differences in levels of genetic introgression with non-native traits and genetic diversity. High levels of non-native variants and genetic variability ($H_O = 0.80$) characterise the Neia River population. Contrarily, lower genetic variability ($H_O = 0.44$) was observed in the Furittu stream population, similarly to what has been observed in other isolated populations on Sardinia [49] and elsewhere (e.g., [35,36]). Additional analysis performed with the colony seems to suggest that the abnormal fish are unrelated. Moreover, in both populations, the observed heterozygosity (H_O; Furittu = 0.49; Neia = 0.80) was similar to the expected heterozygosity (H_E; Furittu = 0.45; Neia = 0.83), suggesting the absence of inbreeding or outbreeding depression. Ultimately, though the small number of markers used does not allow us to exclude that the pugheaded malformation may have a genetic basis, our preliminary analyses suggest inbreeding or outbreeding depression is not likely to be the cause of the observed deformities.

However, we can also exclude the presence of pollutants and pathogens as possible causative factors of this malformation. In fact, the Furittu stream is a well-preserved river with very low anthropogenic pressure. Although sample size was limited, unfavourable abiotic conditions, such as variations in environmental factors such as hypoxia, solar radiation, and temperature during larval development seem to be the most likely factors to trigger the observed deformities. In this context, the Mediterranean streams are often subject to prolonged droughts, which reduce and fragment the trout habitat, with a possible increase in water temperature. The results of this study indicate a need for investigation into the causes and control of head malformation in Mediterranean trout population. Practical conservation management measures should include long-term monitoring programmes in order to estimate the population size and abundance, ecological requirements, and protection of stream habitats. The dissemination of information regarding native trout conservation status and the involvement and education of local people and regional authorities are also crucial for conservation of the species.

Supplementary Materials: The following are available online at http://www.mdpi.com/1424-2818/12/9/353/s1, Table S1: Morphometric data, Upper Jaw Index (UJI) and residuals of sixteen native Mediterranean trout of the Furittu stream (south eastern Sardinia) and one specimen from Neia river. L1. and L2 measurements used to calculate the Upper Jaw Index (UJI), (1) upper jaw depth (DUJ), (2) snout length (LS), (3) orbital horizontal diameter (HO), (4) head depth (HD), (5) orbital vertical diameter (VO), (6) length of maxilla (LM), (7) upper jaw length (LUJ), (8) lower jaw length (LLJ), (9) premaxilla to preoperculum length (LPP), (10) head length at upper jaw (LHU), (11) head length at lower jaw (LHL) and (12) operculum length (LO). Table S2: Parentage analyses performed with COLONY for Riu Furittu population. We have chosen the Pair-Likelihood-Score (PLS)/Full-Likelihood (FL) combined (=FPLS) algorithm and used only full-sibs listing to give the results below.

Author Contributions: Conceptualization, F.P., A.M. and A.S. (Andrea Sabatini); methodology, F.P., A.S. (Andrea Sabatini), A.M., T.R., A.S. (Andrea Splendiani) and V.C.B.; software F.P., A.S. (Andrea Sabatini), A.M., T.R., A.S. (Andrea Splendiani) and V.C.B.; validation, A.S. (Andrea Splendiani), F.P., A.M., C.F., C.P., M.S., T.R., A.S. (Andrea Splendiani) and V.C.B.; formal analysis, F.P., A.S. (Andrea Sabatini)., A.M., T.R., A.S. (Andrea Splendiani) and V.C.B.; investigation, F.P., A.M., C.F., M.S., C.P. and A.S. (Andrea Sabatini); resources, A.S. (Andrea Sabatini); data curation, F.P., A.M., A.S. (Andrea Sabatini), T.R., A.S. (Andrea Splendiani) and V.C.B.; writing—original draft preparation, F.P., A.M., A.S. (Andrea Sabatini), T.R., A.S. (Andrea Splendiani) and V.C.B.; writing—review & editing, F.P., A.M., A.S. (Andrea Sabatini), C.P., T.R., A.S. (Andrea Splendiani) and V.C.B.; visualization, F.P. and A.M.; supervision, A.S. (Andrea Sabatini) and V.C.B.; project administration, A.S. (Andrea Sabatini); funding acquisition, A.S. (Andrea Sabatini). All authors have read and agreed to the published version of the manuscript.

Funding: The research was a part of the Regional Fish Inventory supported by the Regione Sardegna (Redazione di una Carta Ittica delle acque dolci della Sardegna con particolare riferimento ai siti di popolamento della forma geneticamente pura della trota sarda (*Salmo cettii* ex *Salmo trutta macrostigma*), specie autoctona a grave pericolo di estinzione; Grant No. 27002-1 18/12/2015).

Conflicts of Interest: The authors declare no conflict of interest.

References

1. Koumoundouros, G. Morpho-anatomical abnormalities in Mediterranean marine aquaculture. *Recent Adv. Aquac. Res.* **2010**, *661*, 125–148.

2. Boglione, C.; Gavaia, P.; Koumoundouros, G.; Gisbert, E.; Moren, M.; Fontagné, S.; Witten, P.E. Skeletal anomalies in reared European fish larvae and juveniles. Part 1: Normal and anomalous skeletogenic processes. *Rev. Aquac.* **2013**, *5*, 99–120. [CrossRef]

3. Boglione, C.; Gisbert, E.; Gavaia, P.; Witten, P.E.; Moren, M.; Fontagné, S.; Koumoundouros, G. Skeletal anomalies in reared European fish larvae and juveniles. Part 2: Main typologies, occurrences and causative factors. *Rev. Aquac.* **2013**, *5*, 121–167. [CrossRef]

4. Boglione, C.; Costa, C.; Giganti, M.; Cecchetti, M.; Dato, P.; Di Scardi, M.; Cataudella, S. Biological monitoring of wild thicklip grey mullet (*Chelon labrosus*), golden grey mullet (*Liza aurata*), thinlip mullet (*Liza ramada*) and flathead mullet (*Mugil cephalus*) (Pisces: Mugilidae) from different Adriatic sites: Meristic counts and skeletal anoma. *Ecol. Indic.* **2006**, *6*, 712–732. [CrossRef]

5. Jawad, L.A.; Ibrahim, M. Environmental oil pollution: A possible cause for the incidence of ankylosis, kyphosis, lordosis and scoliosis in five fish species collected from the vicinity of Jubail City, Saudi Arabia, Arabian Gulf. *Int. J. Environ. Stud.* **2018**, *75*, 425–442. [CrossRef]

6. Gavaia, P.J.; Domingues, S.; Engrola, S.; Drake, P.; Sarasquete, C.; Dinis, M.T.; Cancela, M.L. Comparing skeletal development of wild and hatchery-reared Senegalese sole (*Solea senegalensis*, Kaup 1858): Evaluation in larval and postlarval stages. *Aquac. Res.* **2009**, *40*, 1585–1593. [CrossRef]

7. Boglione, C.; Pulcini, D.; Scardi, M.; Palamara, E.; Russo, T.; Cataudella, S. Skeletal anomaly monitoring in rainbow trout (*Oncorhynchus mykiss*, Walbaum 1792) reared under different conditions. *PLoS ONE* **2014**, *9*, e96983. [CrossRef]

8. Dawson, C. A bibliography of anomalies of fishes. *Caribb. Res. Rep.* **1964**, *1*, 308–399. [CrossRef]

9. Shariff, M.; Zainuddin, A.T.; Abdullah, H. Pugheadedness in bighead carp, *Aristichthys nobilis* (Richardson). *J. Fish Dis.* **1986**, *9*, 457–460. [CrossRef]

10. Pittman, K.; Bergh, O.I.; Skiftesvik, A.B.; Skjolddal, L.; Strand, H. Development of eggs and yolk sac larvae of halibut (*Hippoglossus hippoglossus* L.). *J. Appl. Ichthyol.* **1990**, *6*, 142–160. [CrossRef]

11. Darias, M.J.; Mazurais, D.; Koumoundouros, G.; Le Gall, M.M.; Huelvan, C.; Desbruyeres, E.; Quazuguel, P.; Cahu, C.L.; Zambonino-Infante, J.L. Imbalanced dietary ascorbic acid alters molecular pathways involved in skeletogenesis of developing European sea bass (*Dicentrarchus labrax*). *Comp. Biochem. Physiol. A Mol. Integr. Physiol.* **2011**, *159*, 46–55. [CrossRef] [PubMed]

12. Schmitt, J.D.; Orth, D.J. First record of pughead deformity in blue catfish. *Trans. Am. Fish. Soc.* **2015**, *144*, 1111–1116. [CrossRef]

13. Slooff, W. Skeletal anomalies in fish from. *Aquat. Toxicol.* **1982**, *2*, 157–173. [CrossRef]

14. Eissa, A.E.; Moustafa, M.; El-Husseiny, I.N.; Saeid, S.; Saleh, O.; Borhan, T. Identification of some skeletal deformities in freshwater teleosts raised in Egyptian aquaculture. *Chemosphere* **2009**, *77*, 419–425. [CrossRef]

15. Jawad, L.A.; Kousha, A.; Sambraus, F.; Fjelldal, P.G. On the record of pug-headedness in cultured Atlantic salmon, *Salmo salar* Linnaeus, 1758 (Salmoniformes, Salmonidae) from Norway. *J. Appl. Ichthyol.* **2014**, *30*, 537–539. [CrossRef]

16. Jawad, L.; Thorsen, A.; Otterå, H.; Fjelldal, P. Pug-headedness in the farmed triploid Atlantic cod, *Gadus morhua* Linnaeus, 1758 (Actinopterygii: Gadiformes: Gadidae) in Norway. *Acta Adriat.* **2015**, *56*, 291–296.

17. Catelani, P.A.; Bauer, A.B.; Di Dario, F.; Pelicice, F.M.; Petry, A.C. First record of pughead deformity in *Cichla kelberi* (Teleostei: Cichlidae), an invasive species in an estuarine system in south-eastern Brazil. *J. Fish Biol.* **2017**, *90*, 2496–2503. [CrossRef]

18. Porta, M.J.; Snow, R.A. First record of pughead deformity in redear sunfish *Lepomis microlophus* (Günther, 1859). *J. Appl. Ichthyol.* **2019**, *35*, 775–778. [CrossRef]

19. Bortone, S.A. Pugheadedness in the Pirate Perch, *Aphredoderus sayanus* (Pisces: Aphredoderidae), with Implications on Feeding. *Chesap. Sci.* **2006**, *13*, 231–232. [CrossRef]

20. Bueno, L.S.; Koenig, C.C.; Hostim-Silva, M. First records of 'pughead' and 'short-tail' skeletal deformities in the Atlantic goliath grouper, *Epinephelus itajara* (Perciformes: Epinephelidae). *Mar. Biodivers. Rec.* **2015**, *8*, 10–13. [CrossRef]

21. Grinstead, B.G. Effects of pigheadedness on growth and survival. *Proc. Okla. Acad. Sci.* **1971**, *12*, 8–12.

22. Francini-Filho, R.B.; Amado-Filho, G.M. First record of pughead skeletal deformity in the queen angelfish *Holacanthus ciliaris* (St. Peter and St. Paul Archipelago, Mid Atlantic Ridge, Brazil). *Coral Reefs* **2013**, *32*, 211. [CrossRef]

23. Franks, J.S. Food of Cobia, *Rachycentron canadum*, from the Northcentral Gulf of Mexico. *Gulf Res. Rep.* **1995**, *9*, 143–145.

24. Jawad, L.; Hosie, A. On the record of pug-headedness in snapper, *Pagrus auratus* (Forster, 1801) (Perciformes, Sparidae) from New Zealand. *Acta Adriat.* **2007**, *48*, 205–210.

25. Jawad, L.A.; Ibrahim, M. Head Deformity in *Epinephelus diacanthus* (Teleostei: Epinephelidae) and *Oreochromis mossambicus* (Teleostei: Cichlidae) Collected from Saudi Arabia and Oman, with a New Record of *E. diacanthus* to the Arabian Gulf Waters. *Thalassas* **2019**, *35*, 263–269. [CrossRef]

26. Gudger, E.W. An adult pug-headed brown trout, *Salmo fario*: With notes on other pug-headed salmonids. *Bull. Am. Mus. Nat. Hist.* **1929**, *58*, 531–564.

27. Splendiani, A.; Palmas, F.; Sabatini, A.; Barucchi, V.C. The name of the trout: Considerations on the taxonomic status of the *Salmo trutta* L., 1758 complex (Osteichthyes: Salmonidae) in Italy. *Eur. Zool. J.* **2019**, *86*, 432–442. [CrossRef]

28. Rondinini, C.; Battistoni, A.; Peronace, V.; Teofili, C. *Lista Rossa IUCN dei Vertebrati Italiani*; IT Stamperia Romana: Rome, Italy, 2013; p. 54. Available online: http://www.iucn.it/liste-rosse-italiane.php (accessed on 1 September 2020).

29. Sabatini, A.; Palmas, F.; Podda, C.; Frau, G.; Serra, M.; Musu, A.; Cani, M.V.; Righi, T.; Splendiani, A.; Caputo Barucchi, V. *Carta Ittica Regionale. Parte 1. Tratti Montani*; Con Appprofondimenti Sulla Distribuzione Della Trota Sarda *Salmo Cettii*, Rafinesque, 1810 Vol. 1. Tecnical Report; Regione Autonoma della Sardegna, Cagliari, Italy, Stamperia Regione Autonoma della Sardegna; 2018; p. 109. Available online: https://portalsardegnasira.it/-/carta-ittica-regionale-parte-i-tratti-momtani (accessed on 1 September 2020).

30. Lockwood, R.N.; Schneider, J.C. Stream fish population esti- mates by mark- and-recapture and depletion methods. In *Manual of Fisheries Survey Methods II: With Periodic Updates*; Schneider, J.C., Ed.; Fisheries Special Report 25; Department for Natural Resources: Ann Arbor, MI, USA, 2000; pp. 1–14.

31. Sušnik, S.; Snoj, A.; Dovč, P. Evolutionary distinctness of grayling (*Thymallus thymallus*) inhabiting the Adriatic river system, as based on mtDNA variation. *Biol. J. Linn. Soc.* **2001**, *74*, 375–385. [CrossRef]

32. Bernatchez, L.; Danzmann, R.G. Congruence in Control-Region Sequence and Restriction-Site Variation in Mitochondrial DNA of Brook Charr (*Salvelinus fontinalis* Mitchill). *Mol. Biol. Evol.* **1993**, *10*, 1002–1014.

33. Thompson, J.D.; Higgins, D.G.; Gibson, T.J. CLUSTAL W: Improving the sensitivity of progressive multiple sequence alignment through sequence weighting, position-specific gap penalties and weight matrix choice. *Nucleic Acids Res.* **1994**, *22*, 4673–4680. [CrossRef]

34. McMeel, O.M.; Hoey, E.M.; Ferguson, A. Partial nucleotide sequences, and routine typing by polymerase chain reaction-restriction fragment length polymorphism, of the brown trout (*Salmo trutta*) lactate dehydrogenase, LDH-C1*90 and *100 alleles. *Mol. Ecol.* **2001**, *10*, 29–34. [CrossRef] [PubMed]

35. Splendiani, A.; Giovannotti, M.; Righi, T.; Fioravanti, T.; Cerioni, P.N.; Lorenzoni, M.; Carosi, A.; La Porta, G.; Barucchi, V.C. Introgression despite protection: The case of native brown trout in Natura 2000 network in Italy. *Conserv. Genet.* **2019**, *20*, 343–356. [CrossRef]

36. Splendiani, A.; Fioravanti, T.; Ruggeri, P.; Giovannotti, M.; Carosi, A.; Marconi, M.; Righi, T.; Nisi Cerioni, P.; Lorenzoni, M.; Caputo Barucchi, V. Life history and genetic characterisation of sea trout Salmo trutta in the Adriatic Sea. *Freshw. Biol.* **2020**, *65*, 460–473. [CrossRef]

37. Excoffier, L.; Lisher, H.E.L. ARLEQUIN suite ver 3.5: A new series of programs to perform population genetics analyses under Linux and Windows. *Mol. Ecol. Res.* **2010**, *10*, 564–567. [CrossRef] [PubMed]

38. Goudet, J. *FSTAT, a Program to Estimate and Test Gene Diversities and Fixation Indices*, version 2.9. 3; Institute of Ecology: Lausanne, Switzerland, 2001.

39. Jones, O.R.; Wang, J. COLONY: A program for parentage and sibship inference from multilocus genotype data. *Mol. Ecol. Res.* **2010**, *10*, 551–555. [CrossRef]

40. Pritchard, J.K.; Stephens, M.; Donnelly, P. Inference of population structure using multilocus genotype data. *Genetics* **2000**, *155*, 945–959.

41. Lijalad, M.; Powell, M.D. Effects of lower jaw deformity on swimming performance and recovery from exhaustive exercise in triploid and diploid Atlantic salmon *Salmo salar* L. *Aquaculture* **2009**, *290*, 145–154. [CrossRef]

42. Delling, B.; Crivelli, A.J.; Rubin, J.F.; Berrebi, P. Morphological variation in hybrids between *Salmo marmoratus*, and alien *Salmo* species in the Volarja stream, Soca River basin, Slovenia. *J. Fish. Biol.* **2000**, *57*, 1199–1212.

43. Rohlf, F.J. The tps series of software. *Hystrix* **2015**, *26*, 1–4.

44. Bernatchez, L. The Evolutionary History of Brown Trout (*Salmo Trutta* L.) Inferred From Phylogeographic, Nested Clade, and Mismatch Analyses of Mitochondrial Dna Variation. *Evolution* **2001**, *55*, 351–379. [CrossRef]

45. Berra, T.; Au, R.J. Incidence of Teratological Fishes from Cedar Fork Creek, Ohio. *Ohio J.* **1981**, *81*, 225–229.

46. Abdel, I.; Abellan, E.; Lopez-Albors, O.; Valdés, P.; Nortes, M.J.; Garcia-Alcazar, A. Abnormalities in the juvenile stage of sea bass (*Dicentrarchus labrax* L.) reared at different temperatures: Types, prevalence and effect on growth. *Aquac. Int.* **2004**, *12*, 523–538. [CrossRef]

47. Sabatini, A.; Cannas, R.; Follesa, M.C.; Palmas, F.; Manunza, A.; Matta, G.; Pendugiu, A.A.; Serra, P.; Cau, A. Genetic characterization and artificial reproduction attempt of endemic Sardinian trout *Salmo trutta* L., 1758 (Osteichthyes, Salmonidae): Experiences in captivity. *Ital. J. Zool.* **2011**, *78*, 20–26. [CrossRef]

48. Sabatini, A.; Podda, C.; Frau, G.; Cani, M.V.; Musu, A.; Serra, M.; Palmas, F. Restoration of native Mediterranean brown trout *Salmo cettii* Rafinesque, 1810 (Actinopterygii: Salmonidae) populations using an electric barrier as a mitigation tool. *Eur. Zool. J.* **2018**, *85*, 138–150. [CrossRef]

49. Zaccara, S.; Trasforini, S.; Antognazza, C.M.; Puzzi, C.; Britton, J.R.; Crosa, G. Morphological and genetic characterization of Sardinian trout *Salmo cettii* Rafinesque, 1810 and their conservation implications. *Hydrobiologia* **2015**, *760*, 205–223. [CrossRef]

50. Berrebi, P.; Caputo Barucchi, V.; Splendiani, A.; Muracciole, S.; Sabatini, A.; Palmas, F.; Tougard, C.; Arculeo, M.; Marić, S. Brown trout (*Salmo trutta* L.) high genetic diversity around the Tyrrhenian Sea as revealed by nuclear and mitochondrial markers. *Hydrobiologia* **2018**, *826*, 209–231. [CrossRef]

51. Delling, B.; Sabatini, A.; Muracciole, S.; Tougard, C.; Berrebi, P. Morphologic and genetic characterisation of Corsican and Sardinian trout with comments on *Salmo* taxonomy. *Knowl. Manag. Aquat. Ecosyst.* **2020**, *421*, 21. [CrossRef]

52. Bruno, D.W.; Noguera, P.A.; Poppe, T.T. *A Color Atlas Salmonid Diseases*; Springer: Dordrecht, The Netherlands; Heidelberg, Germany; New York, NY, USA; London, UK, 2013; p. 220.

53. Damsgard, B.; Juell, J.; Braastad, B. *Welfare in Farmed Fish*; Report 5; Fiskeriforskning: Bergen, Norway, 2006.

54. Villeneuve, L.; Gisbert, E.; Zambonino-Infante, J.L.; Quazuguel, P.; Cahu, C.L. Effect of nature of dietary lipids on European sea bass morphogenesis: Implication of retinoid receptors. *Br. J. Nutr.* **2005**, *94*, 877–884. [CrossRef]
55. Peruzzi, S.; Puvanendran, V.; Riesen, G.; Seim, R.R.; Hagen, Ø.; Martínez-Llorens, S.; Falk-Petersen, I.B.; Fernandes, J.M.O.; Jobling, M. Growth and development of skeletal anomalies in diploid and triploid Atlantic salmon (*Salmo salar*) fed phosphorus-rich diets with fish meal and hydrolyzed fish protein. *PLoS ONE* **2018**, *13*, e0194340. [CrossRef]
56. Sato, T.; Harada, Y. Loss of genetic variation and effective population size of Kirikuchi charr: Implications for the management of small, isolated salmonid populations. *Anim. Conserv.* **2008**, *11*, 153–159. [CrossRef]
57. Vincenzi, S.; Crivelli, A.J.; Jesenšek, D.; De Leo, G.A. The management of small, isolated salmonid populations: Do we have to fix it if it ain't broken? *Anim. Conserv.* **2010**, *13*, 21–23. [CrossRef]
58. Morita, K.; Yamamoto, S. Occurrence of a deformed white-spotted charr, *Salvelinus leucomaenis* (Pallas), population on the edge of its distribution. *Fish. Manag. Ecol.* **2000**, *7*, 551–553. [CrossRef]
59. Sato, S. Occurrence of Deformed Fish and Their Fitness-related Traits in Kirikuchi Charr, *Salvelinus ieucomaenis japonicus*, the Southernmost Population of the Genus *Salvelinus*. *Zool. Sci.* **2006**, *23*, 593–599. [CrossRef] [PubMed]

MDPI

St. Alban-Anlage 66

4052 Basel

Switzerland

Tel. +41 61 683 77 34

Fax +41 61 302 89 18

www.mdpi.com

Diversity Editorial Office

E-mail: diversity@mdpi.com

www.mdpi.com/journal/diversity